ChemScrapery

A cartoon collection from the world of ChemScrapes

Brendan A Burkett

ChemScrapery: A Cartoon Collection from the World of ChemScrapes

Published by: Brendan Adrian Burkett

www.chemscrapes.net

Follow ChemScrapes on:

Twitter: @ChemScrapes
Facebook: ChemScrapes
Instagram: @Chem_Scrapes

ISBN: 9789811172267

For Misty, Amelia and Vincent

Contents

ACKNOWLEDGEMENTS

The journey that has led to putting this collection together has indeed been a long one, and there are numerous people who have helped along the way who I would like to acknowledge. An extra special thanks is due to Professor Brian Halton for his generous time, good humour and suggestion a proper title was needed as a home for my growing collection of cartoons. His thoughtfulness in coming up with *"ChemScrapes"* as a title and for giving me a run in *Chemistry in New Zealand* is greatly appreciated.

I would like to acknowledge Christina Chai who, nearly twenty years ago, put my name forward to the editors of *Chemical Innovation* as a possible contributor of cartoons. I am also indebted to Michael Block and Julia Belcher from *Chemical Innovation* for their responsiveness and for giving me the opportunity to publish those early cartoons.

Thanks to Mitch Andre Garcia for patiently reinstating me when I kept getting barred from *Chemistry Reddit,* and for adding *ChemScrapes* to the blogroll of the *Chemistry Blog*, giving some very much needed exposure. More recently, a big, big thanks to @Chemjobber, @AlexFGoldberg, @EdSherer, @RealTimeChem and @SuperScienceGrl for the enormous generosity shown as *ChemScrapes* bumbled into the world of social media. I am also grateful to Amanda Yarnell and Lauren Wolf from *Chemical and Engineering News* for making *Sketch Chemistry* a reality, and especially to Michael Torrice who is a fantastic collaborator on the panel.

On the home front, I am grateful to my brother, Harry, for taking the time to proofread and for his very insightful feedback and suggestions. Finally, to Misty, Amelia and Vincent for their endless patience, curiosity, good humour and encouragement as I pursue my combined loves of chemistry, drawing and dad jokes.

Introduction

ChemScrapes has been around in the public domain in one form or another for nearly twenty years, beginning in the late 1990s with a few terrible drawings appearing on the front and back covers of *RSC News* (the newsletter for Research School of Chemistry at the Australian National University). Early cartoons were, for the most part, terribly drawn and suffered from an inability to bring any depth to the situations, whilst focussing on providing too much detail (e.g. insisting every hand had five fingers *with* fingernails showing). At the time I struggled immensely with drawing people, so most early efforts were focussed on bad chemistry-related puns and bringing personalities to glassware, molecules and atoms—all of which now form the backbone of the current *ChemScrapes* repertoire. Between 2000 and 2001, several cartoons were published by the American Chemical Society publication *Chemical Innovation*, all of which are still available online (http://pubs.acs.org/subscribe/archive/ci/index.html). Probably half of the cartoons submitted to *Chemical Innovation* were rejected, many of which make their first public appearance on the following pages.

"Steve—we've had complaints. Apparently you are getting too worked up in the lab."

Worked-up (2000)

"Dunno what's going on. They just showed up after the opening ceremony"

Olah-la(2000)

Kekule and his less talked about dream about snakes

Kekule (2000)

Argon (2000)

"Dave, this is not what I meant when I told you to sort out your problem reaction!"

Gloves off (2000)

Aldol condensation (2000)

Iron Pyrite (2001)

"Well, there goes Harry. But what's that driving him?"

Driving Force (2000)

A Reducing Agent (2000)

Propyne (2000)

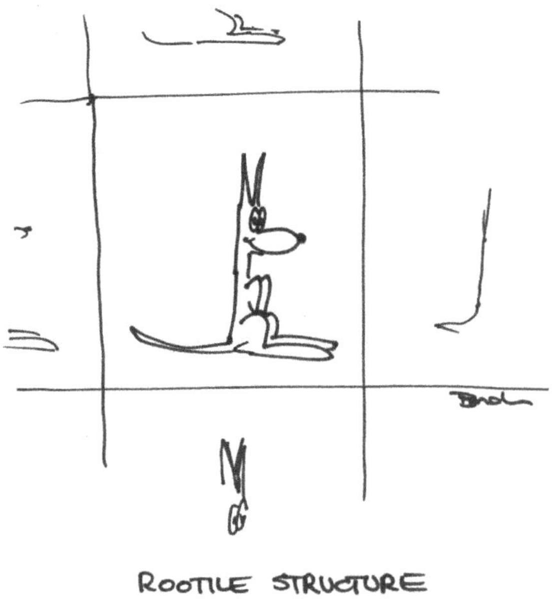

Rutile Structure (2000)

When *Chemical Innovation* ceased operation at the end of 2001, the first attempts were made to establish a presence on the web, first through *Chemical and Scientific Cartoons* and later through *Chempilations*. Neither of these efforts were very successful as I had no idea how to reach my target audience, but the activity at least allowed me to develop some basic skills in web design.

By the time I started my academic career at the School of Chemical and Physical Sciences at Victoria University of Wellington at the end of 2004, I had basically given up on the idea of publishing more cartoons and turned my efforts to using cartoons to support my teaching. My neighbour in the School for the first couple of years was Emeritus Professor Brian Halton who encouraged me to keep going, convincing me all I needed was to give the cartoons a home under a catchy title. After some discussion he proposed the name *ChemScrapes* as a counter suggestion to my own, *ChemScapes*. I instantly fell in love with his idea, because it implied somehow that everything was substandard: the humour, the artwork...***everything!*** In 2008 he allowed me some space in *Chemistry in New Zealand*, which he was editor of at the time, and *ChemScrapes* was officially born.

2003

2004

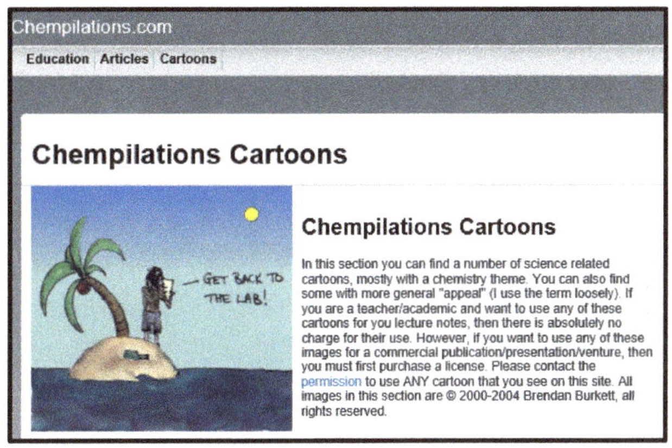

2010

2012

2016

The evolution of ChemScrapes on the web

After leaving academia in 2008, I returned to the internet and was stunned at how much had changed in the four year gap between websites - things had moved on a lot. Facebook and Twitter had appeared, and there was a substantial Blogging community which I previously had no idea about. I started promoting cartoons through *Chemistry Reddit* with mixed success, but it took another six years of running a blog and experimenting on social media to start to get any level of interaction. In 2016 I *finally* asked somebody at work what hashtags were all about and the penny dropped. For the first time I was able to reach my target audience, allowing me to interact in real time with fellow chemists. As a result, *ChemScrapes* was lucky to get two mentions on *Chemical and Engineering News'* list of "Twitter moments of the year for 2016". Then, at the end of 2016, *Chemical and Engineering News* approached me to try out a new cartoon collaboration for their printed and online platforms, and in January 2017 *Sketch Chemistry* made its first appearance as a monthly panel.

Recently, I realised there are nearly 500 cartoons from the last 19 years scattered in various folders on my laptop. Over this time period, cartoons have been produced using pen, pencil, scrap paper, cartridge paper, watercolours, scanners, mobile phone cameras and more recently, drawing tablets with proper graphic-art software. The drawing style has also changed significantly over the years. Given all this change, I thought it would be interesting to capture the evolution of *ChemScrapes* by bringing a selection of cartoons together into print for the first time. The resulting collection contains nearly 200 images, including full colour revamps and modernised versions of some of the older material produced especially for this book.

On reflection, whilst there have been changes in style and substance over the years, it is clear that my passion for chemistry has not waned. Ultimately, it is this passion that is the driving force behind *ChemScrapes*. If you are reading this book you are also clearly passionate about chemistry (or a supportive family member), and my hope is that you face-palm, shake your head, smile, chuckle and groan your way through the pages as they scrape the bottom of the chemical barrel.

Happy *ChemScraping*!

"Cartoon Space" (2002, still from hand drawn animation)

(http://www.chemscrapes.net/single-post/2018/05/04/An-Old-Animation)

The Molecular Level

Tetrahydrofurious (2000)

Vampyran (2001)

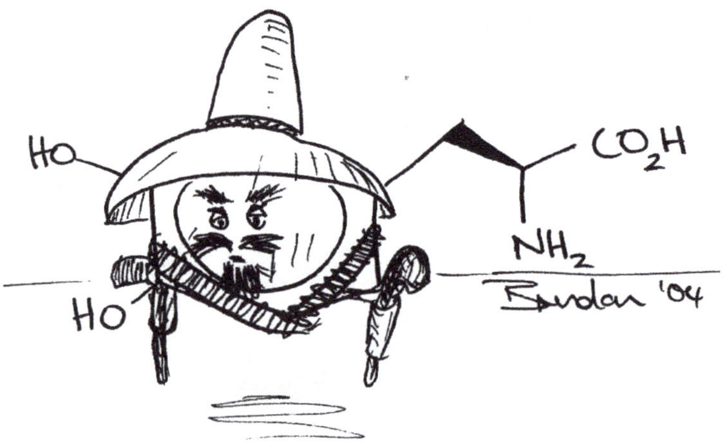

El Dopa (2003)

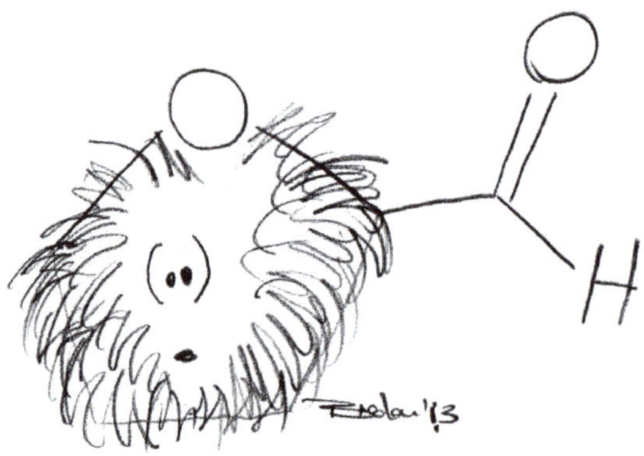

Furfuraldehyde (2013)

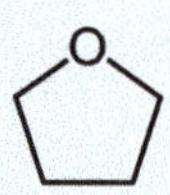

Tetrahydrofuran

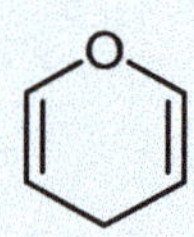

Pyran

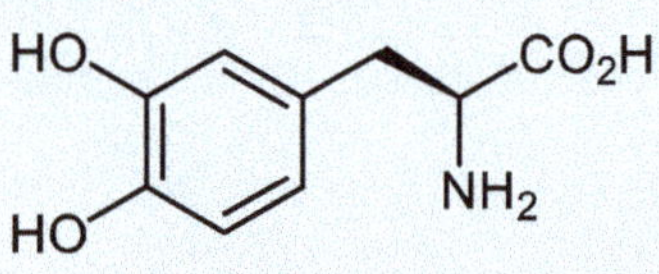

L-*dopa*

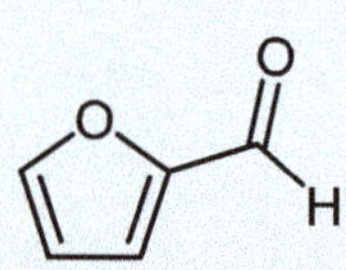

Furfuraldehyde

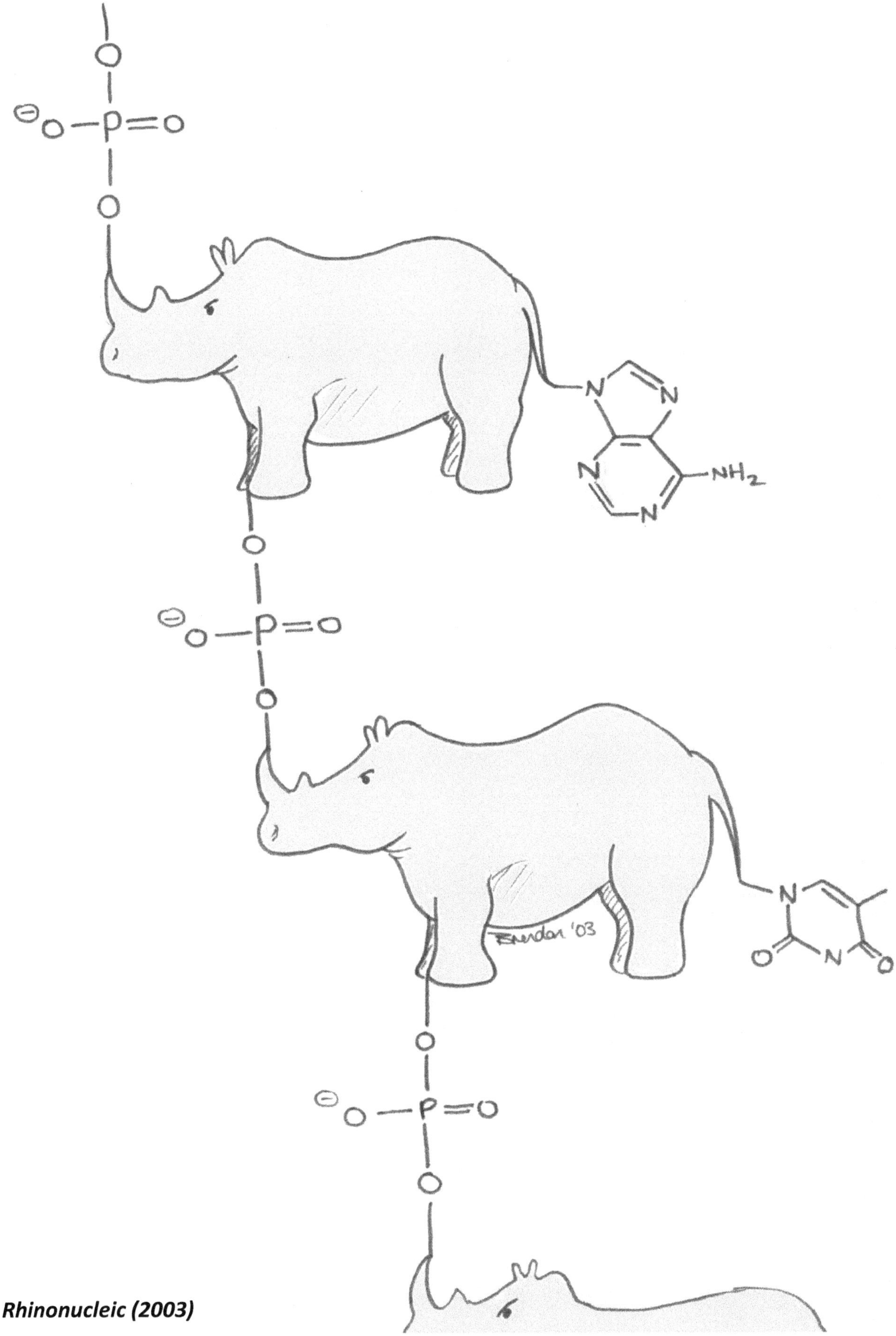

RNA: Rhinonucleic (2003)

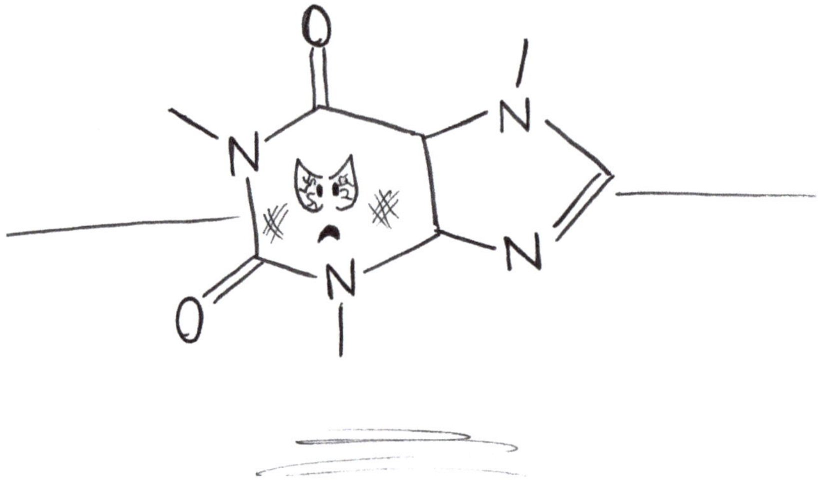

"Don't even think of talking to me until I have had my coffee!"

Morning coffee (2003)

"No thanks. Those things will kill ya!"

Smoke-free (2003)

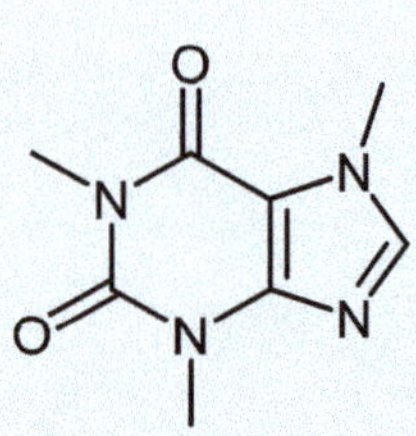

Caffeine

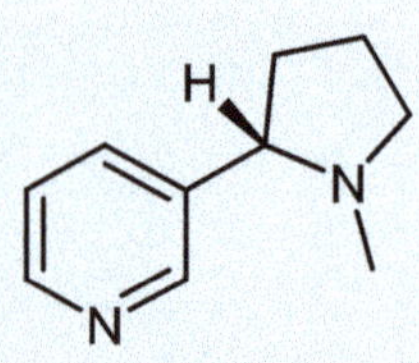

Nicotine

Water Skiing (2003)

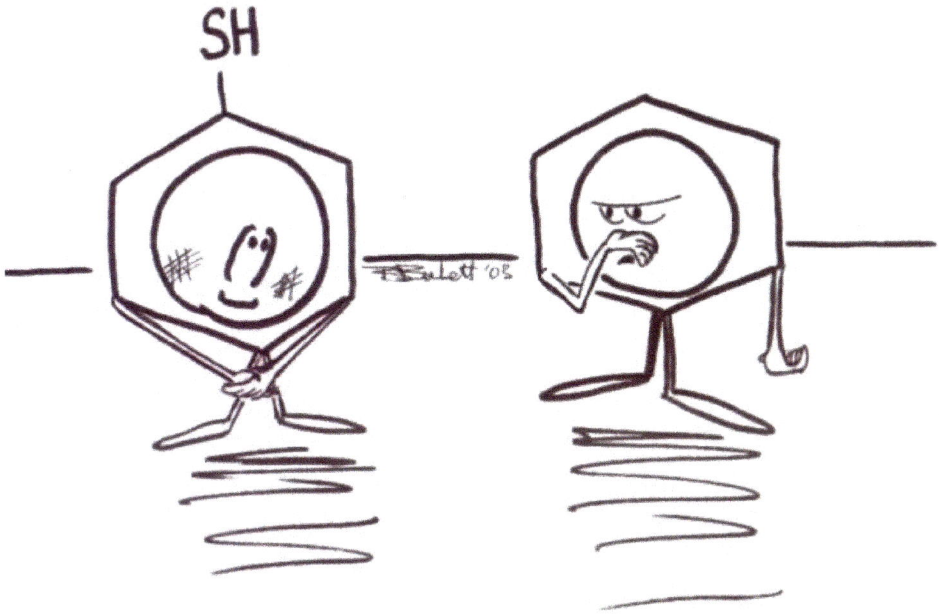

"Pooooooh-Weeee! Was that you?"

Thiol problems (2003)

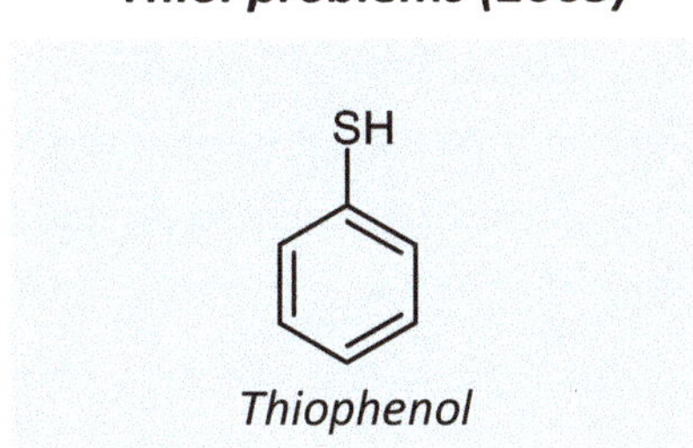

SN1: The rate determining step (2004)

This is amongst the most popular of all the ChemScrapes cartoons that has ever appeared online, and was drawn when I was preparing for teaching mechanistic chemistry at Victoria University of Wellington. It turned out to be a very engaging teaching tool and I frequently had students come up to me after lectures suggesting that I should have more starting material waiting to go up the rate-determining-step, or perhaps the molecules should be running after losing the bromine, or going down a slide once they get to the top. I was always pleased with such comments, because it had students talking about the underpinning theories related to the cartoon.

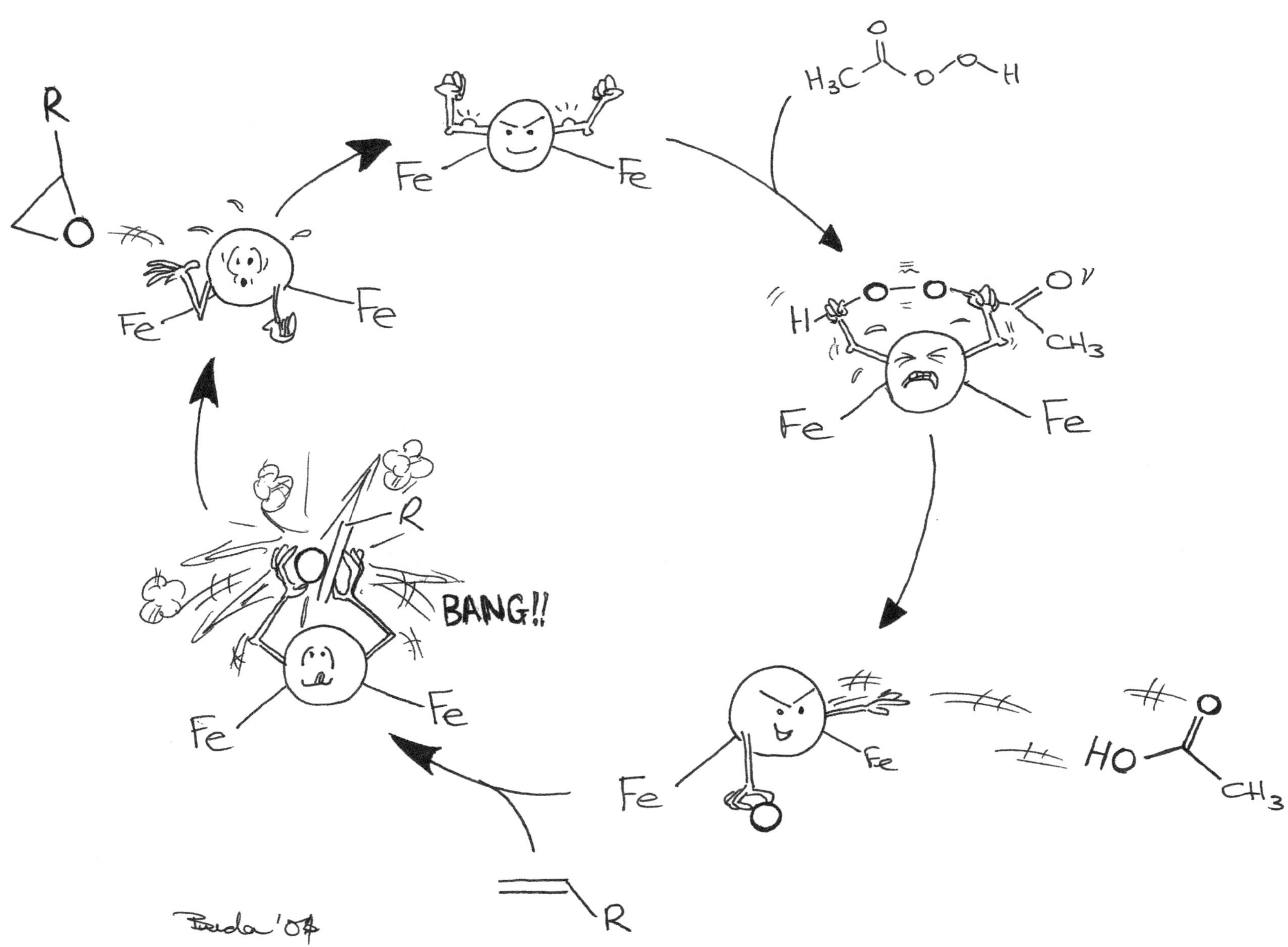

μ-oxo-bridged iron dimer catalysed epoxidation of olefins (2004)

"Thanks Doc, I'll give you a call in the morning and let you know how I feel."

Take two aspirin (2003)

Pyridine blind dates (2003)

Pyridine *Aspirin*

Much to his embarrassment, after all his boasting at the pub, Franky Fluorescein was unable to reach an excited state.

Failed states (2004)

Franky Fluorescein fails to achieve an excited state.

Failed states (2018)

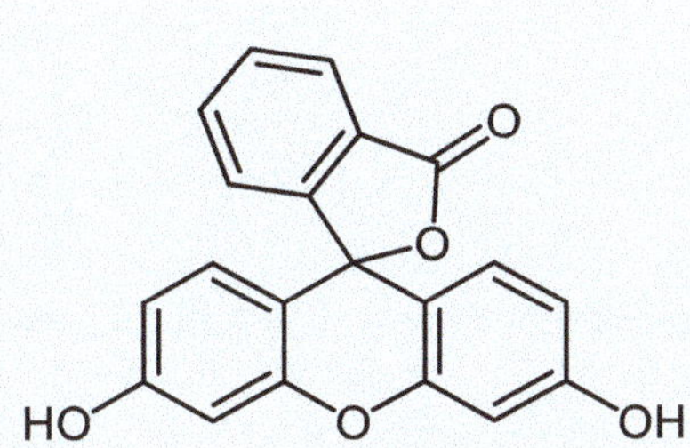

Fluorescein

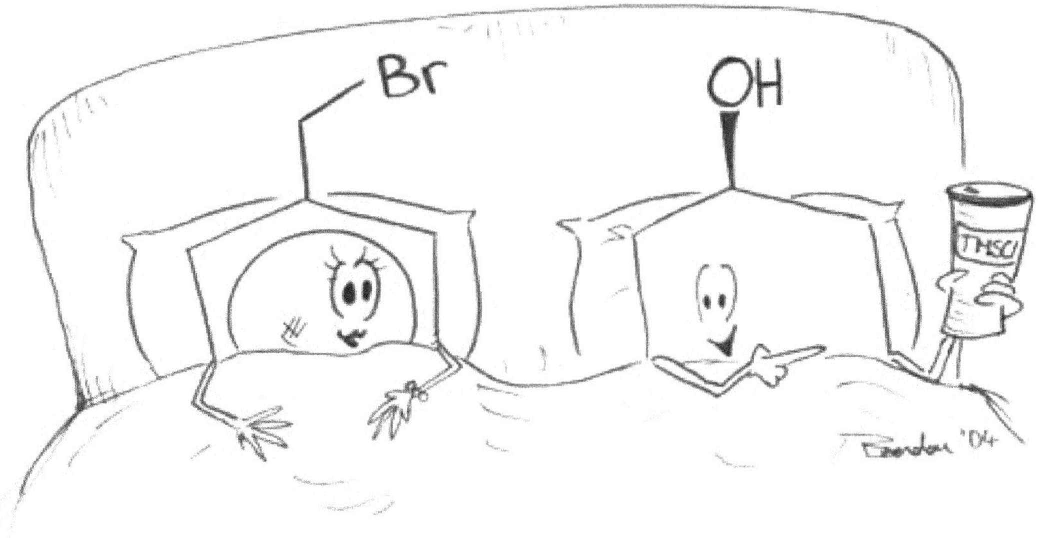

"Don't worry….I brought protection!"

Protecting groups (2004)

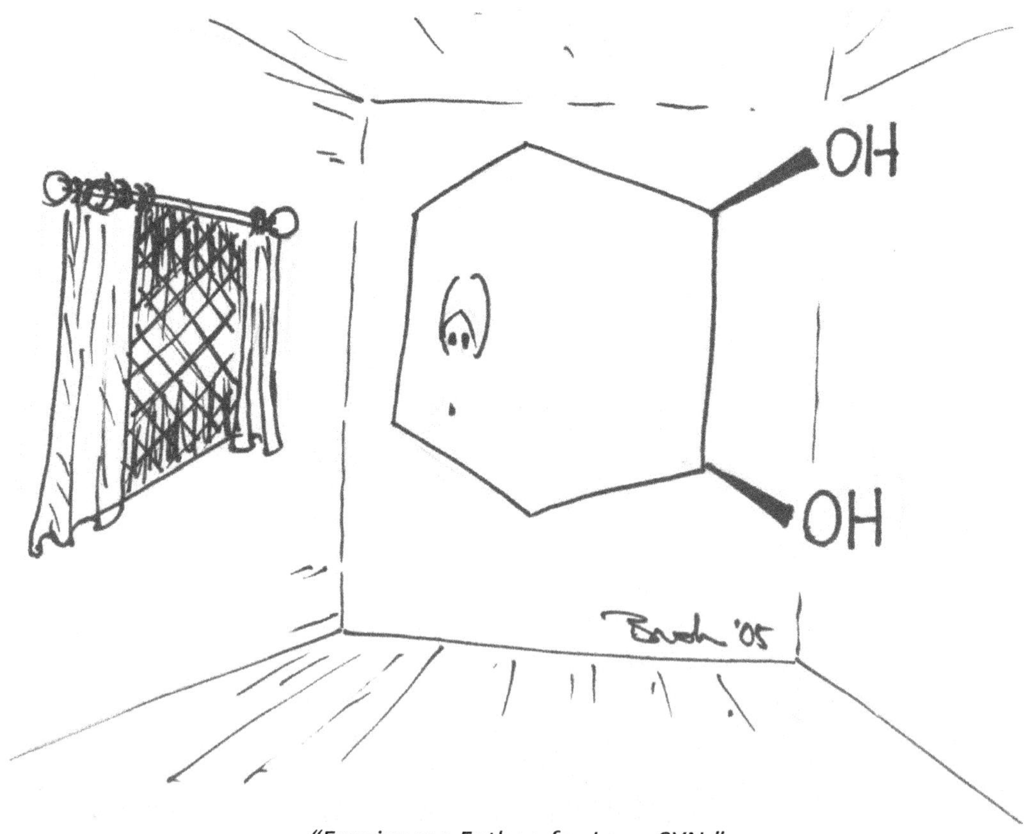

"Forgive me Father, for I am SYN."

Molecular confessionals (2005)

Supersaturated (2010)

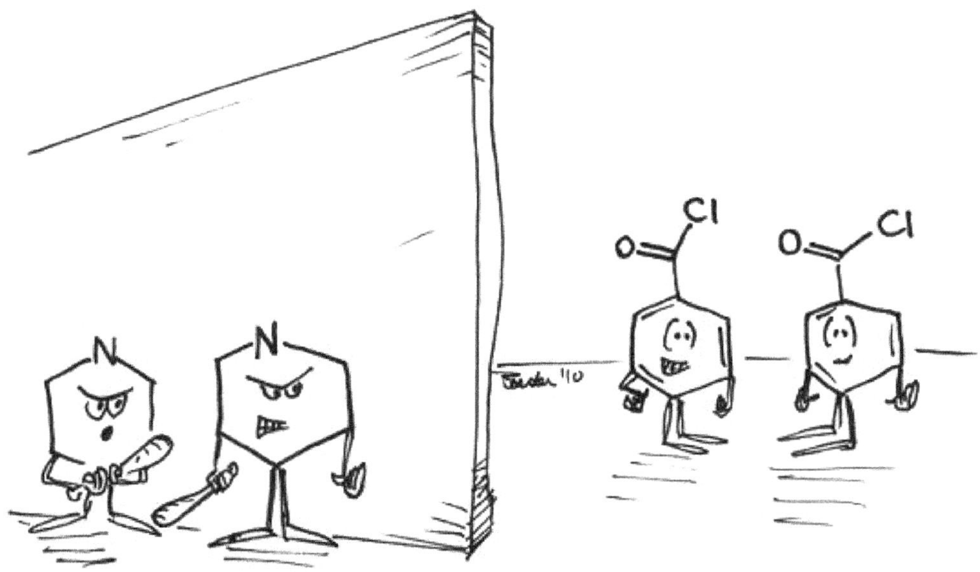

"OK. On the count of three we get 'em."

Nucleophilic attacks (2010)

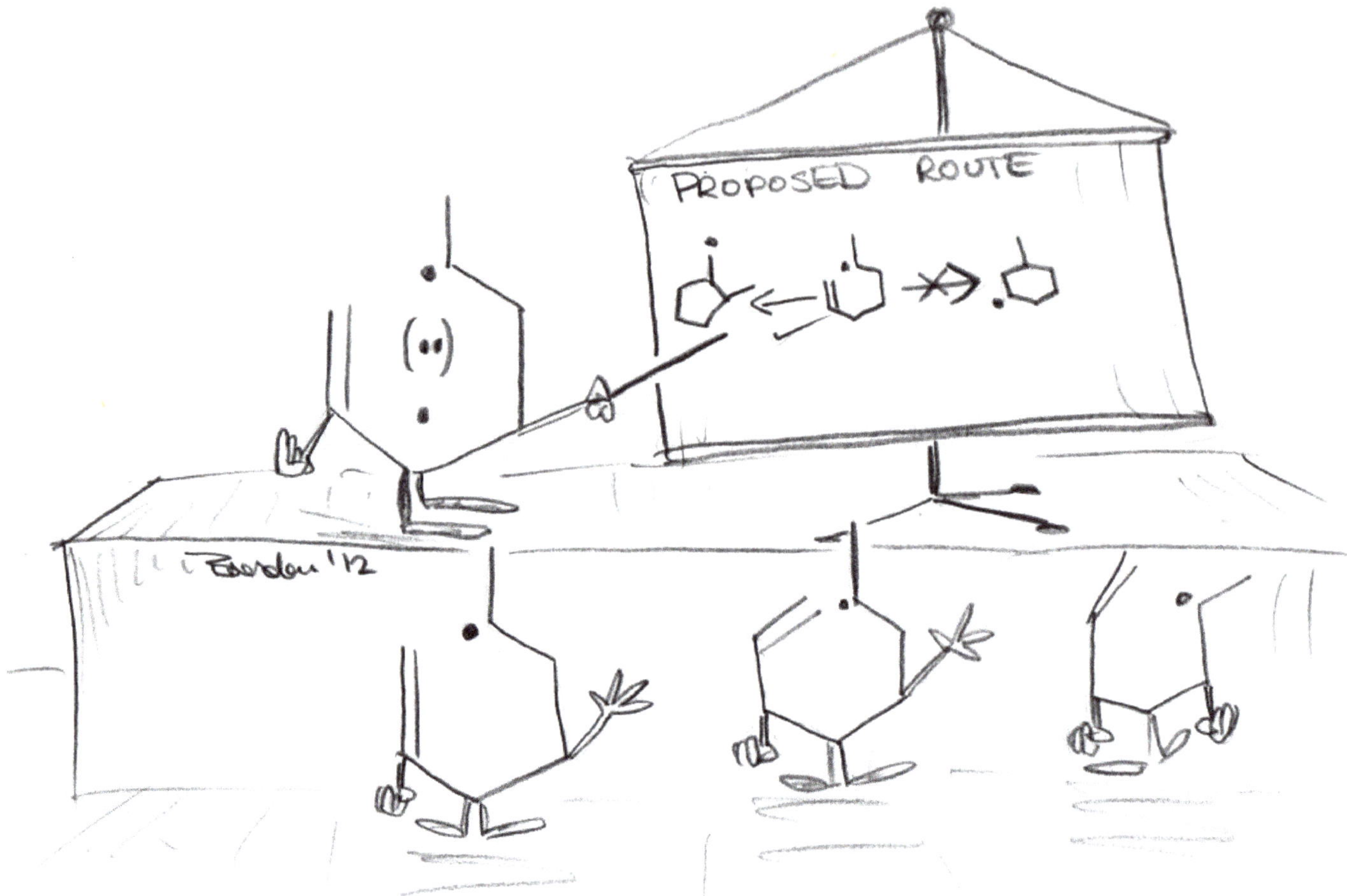

"Those in favour?"

Baldwin-Beckwith (2012)

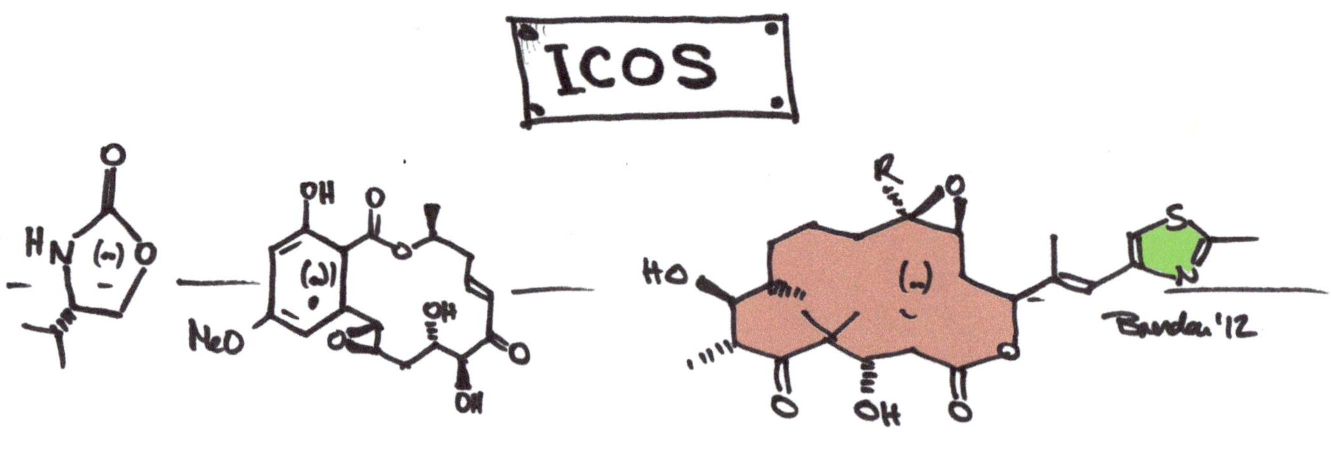

"Must be another one from the Nicolaou lab."

Colouring-in (2012)

Solvent cages (2012)

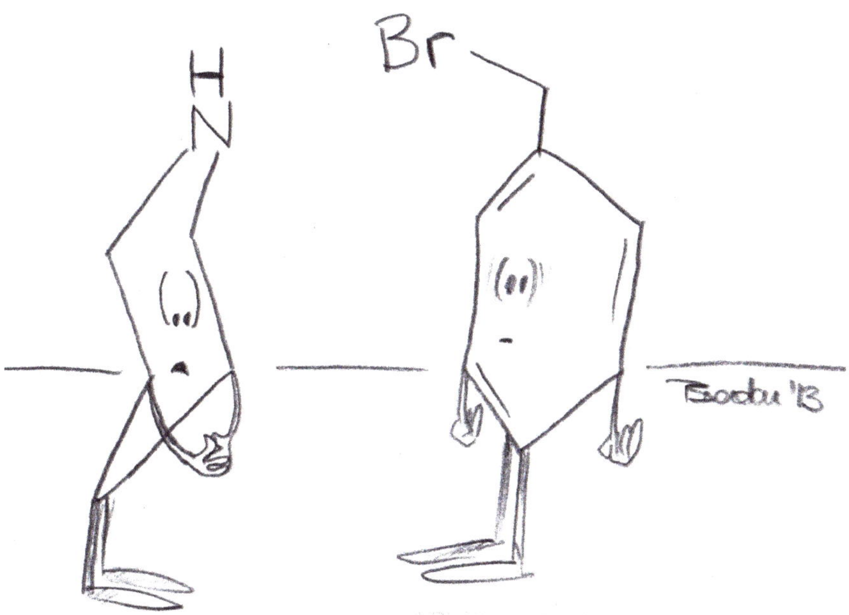

"I know we have only just met, and this may seem forward. Would you mind turning around for just a bit?"

SN2: The rate determining step (2013)

Drug testing (2013)

Aspirin and Ibuprofen (2013)

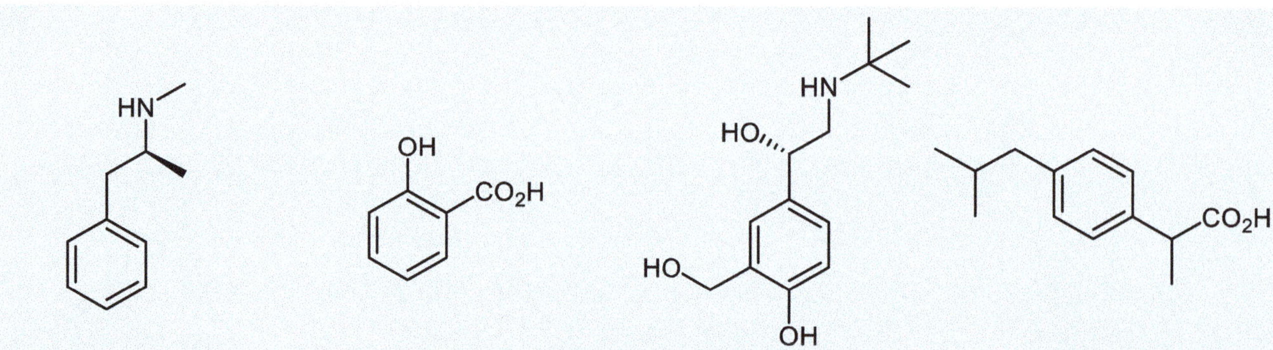

Methamphetamine *Salicylic acid* *Salbutamol* *Ibuprofen*

"ME? I distinctly remember you saying that YOU would bring the palladium!"

Metal free, cross, coupling (2015)

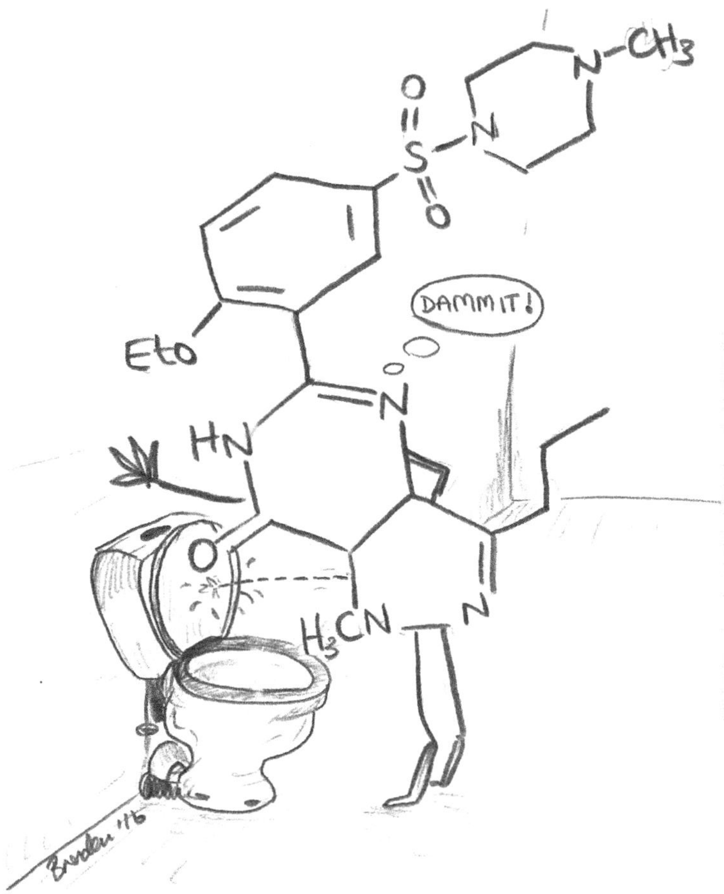

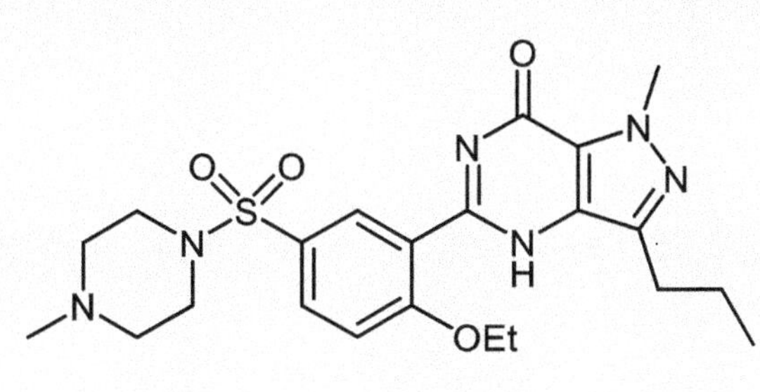

Viagra (Sildenafil)

The curse of Viagra (2015)

Spin trap (2015)

Raising the TEMPO (2016)

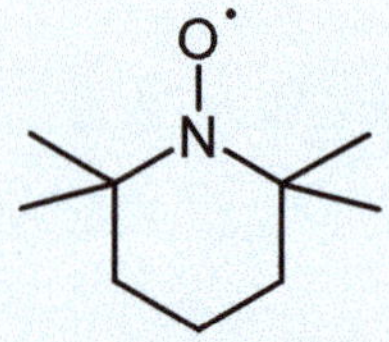

TEMPO
(tetramethylpiperidine-N-oxide)

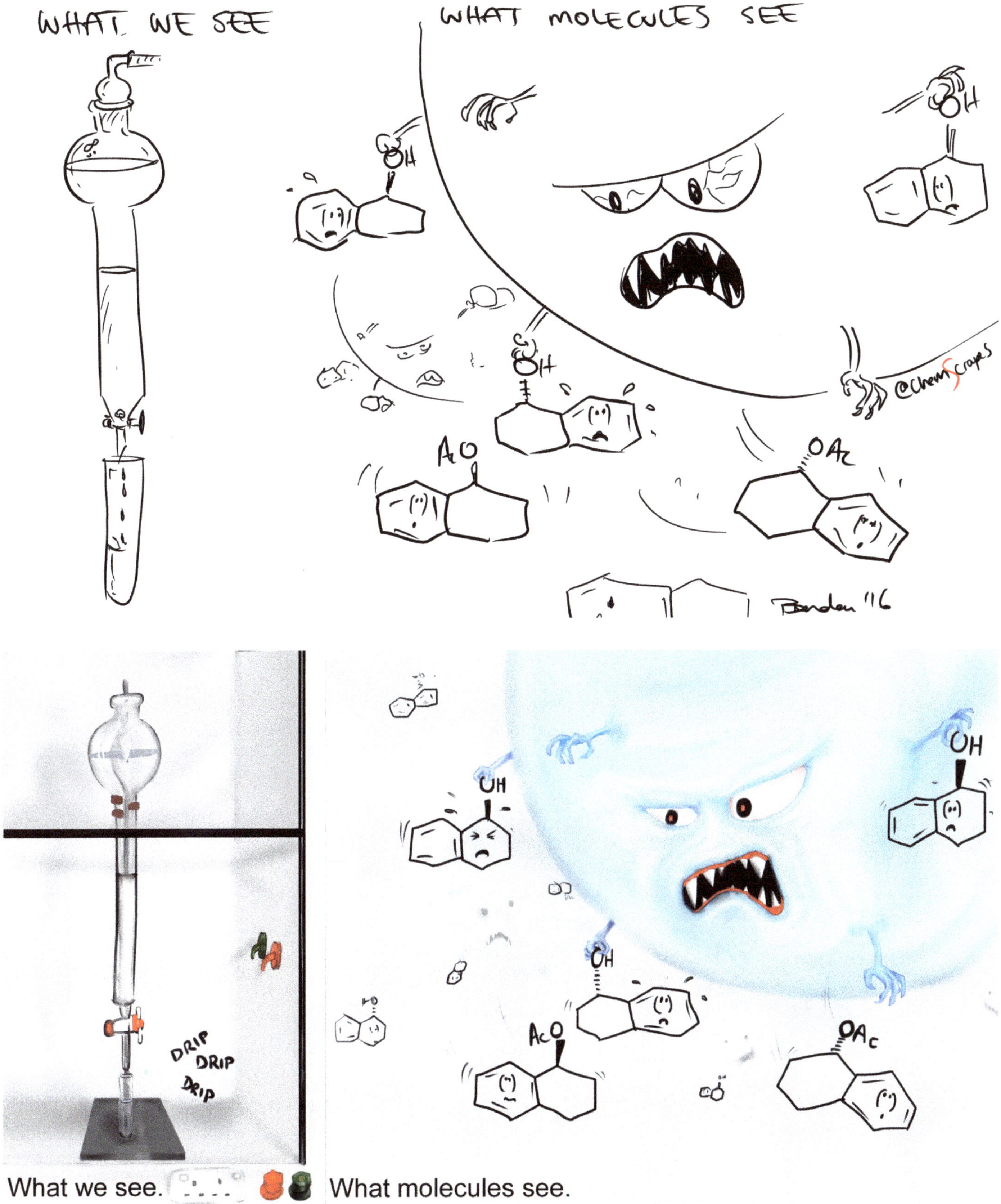

Chromatography: a matter of perspective (2016 (top), 2018 (bottom))

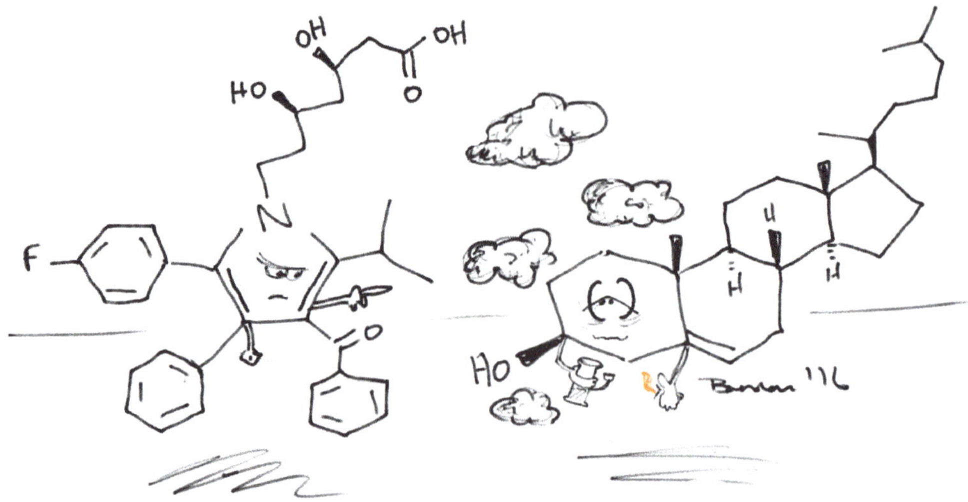

Lipitor regulating high cholesterol (2016)

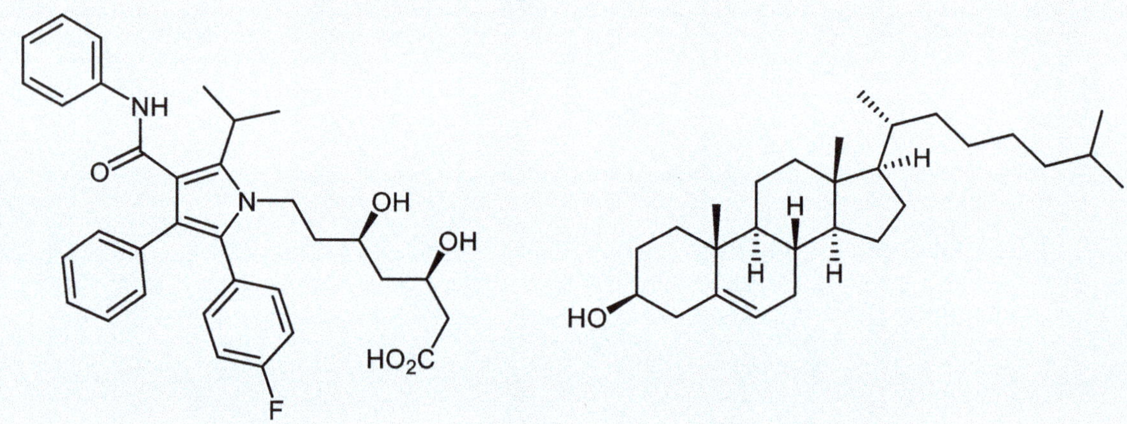

Lipitor *Cholesterol*

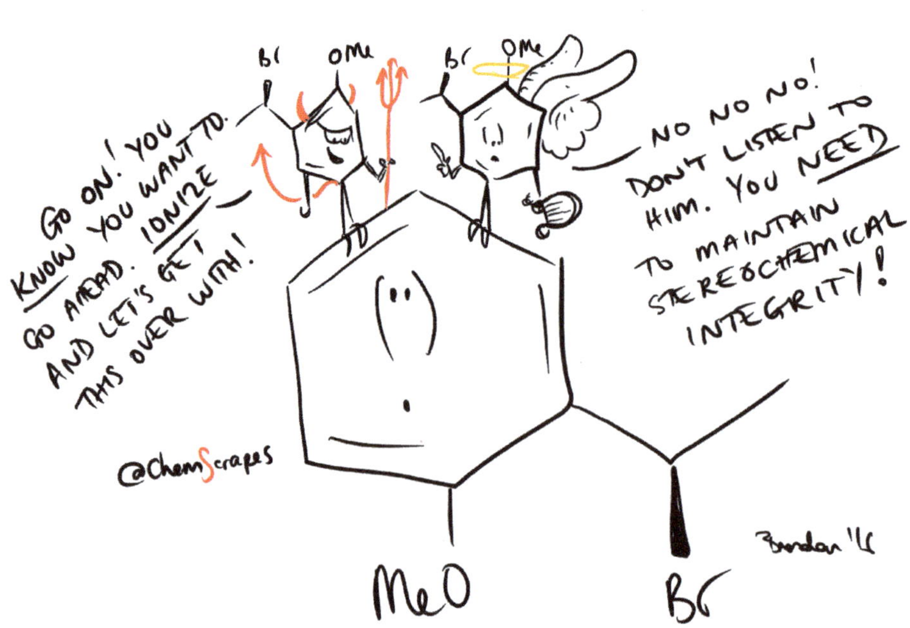

At the mechanistic crossover (2016)

Dud gifts (2016)

Heavy water (2016)

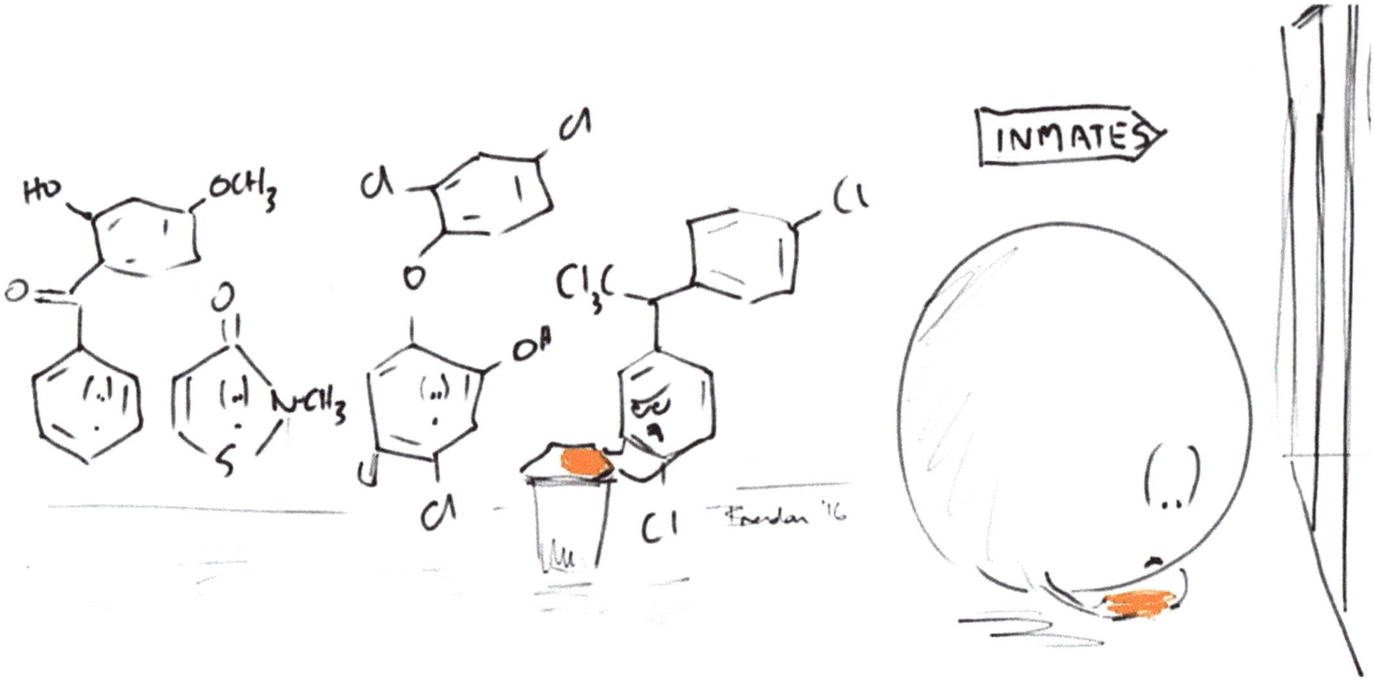

*"I don't care if orange isn't your color. Now, take the *#$@ uniform, get changed and report to the yard!"*

Banned (2016)

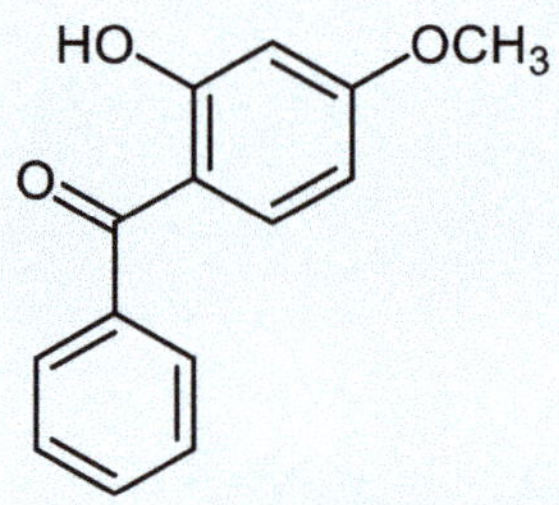

Oxybenzone
(UV filter in sunscreens)

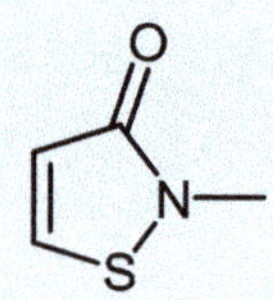

Methyl isothioazolinone
(preservative)

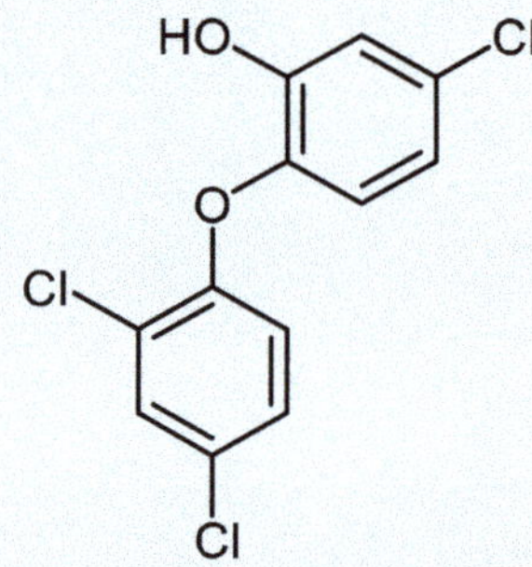

Triclosan
(antiseptic additive)

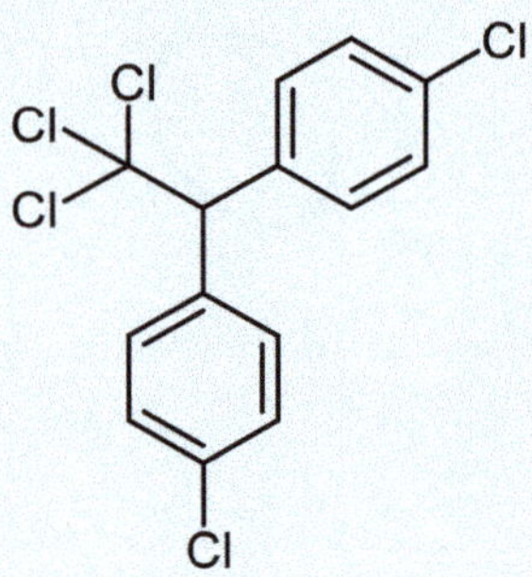

Dichlorodiphenyltrichloroethane (DDT)
(pesticide)

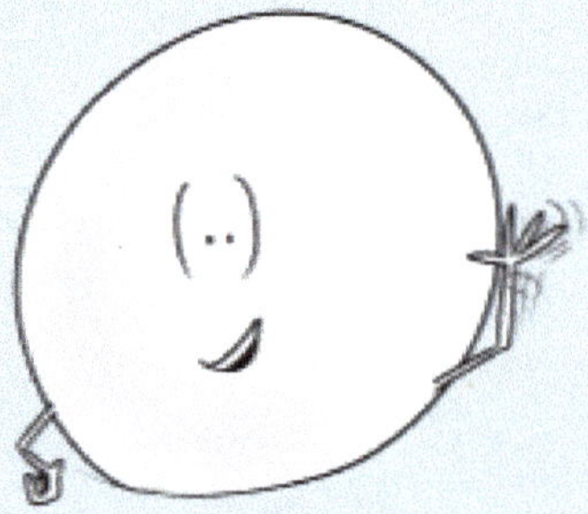

Microbeads
(exfoliants)

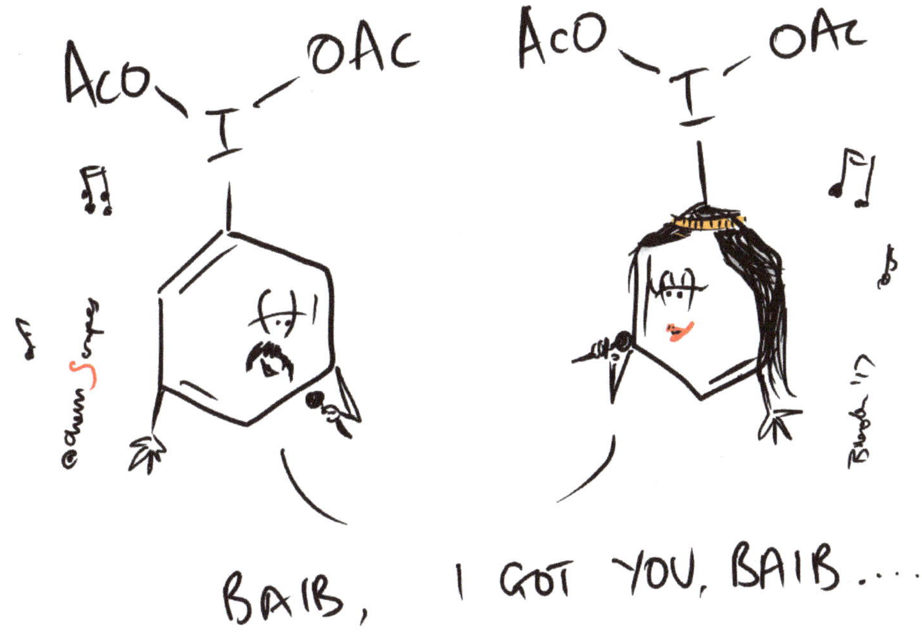

Sonny and Cher (2017)

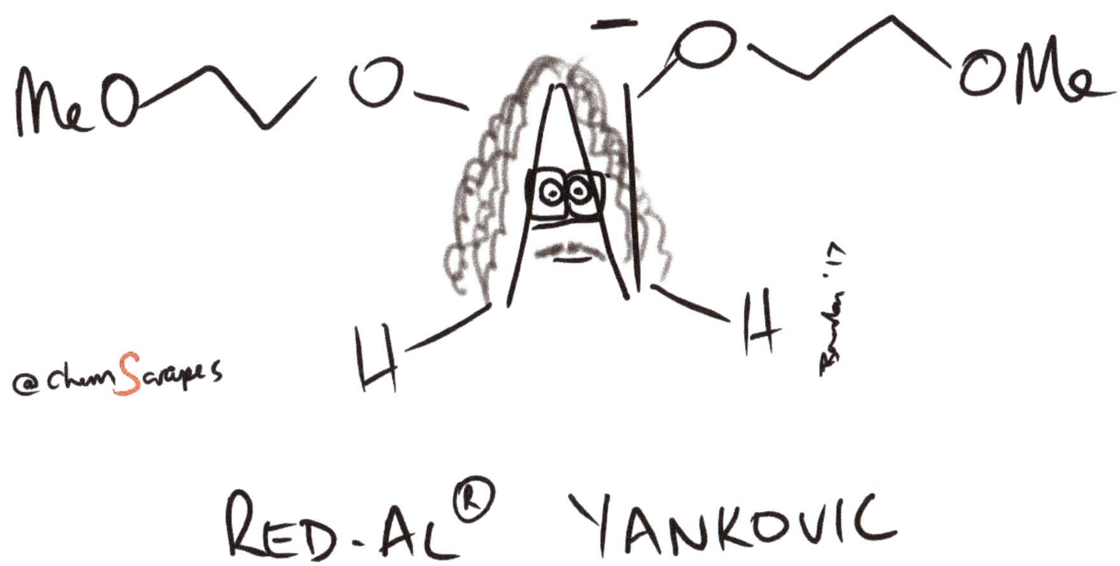

Red Al (2017)

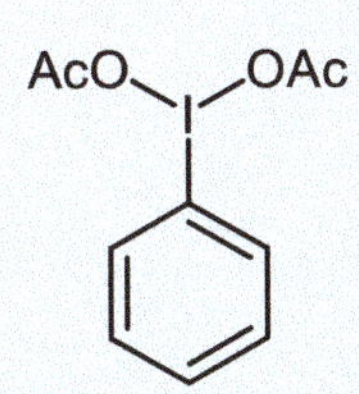

Bis-acetoxyiodobenzene (BAIB)

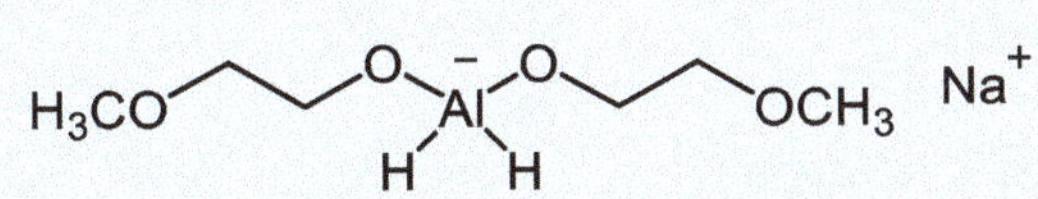

Red-Al®
Sodium bis(2-methoxyethoxy)aluminium hydride

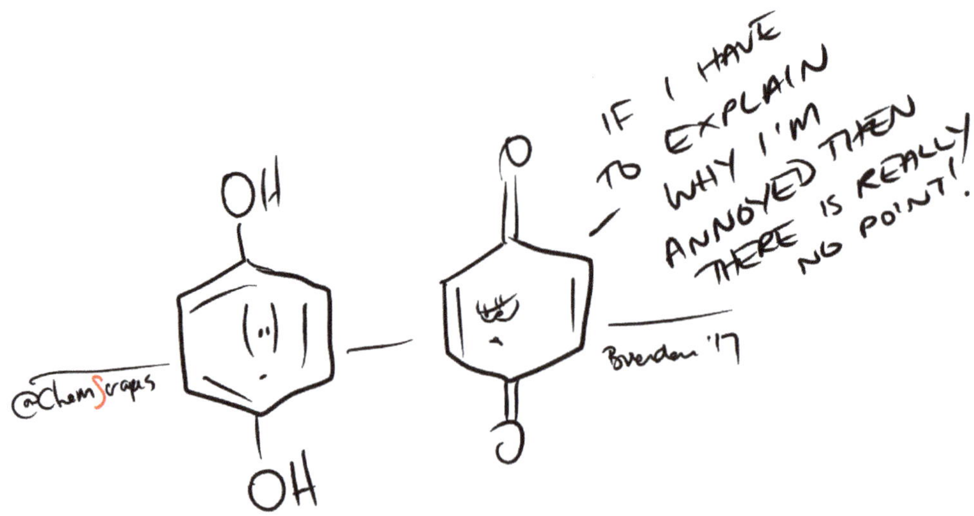

Redox couples (2017)

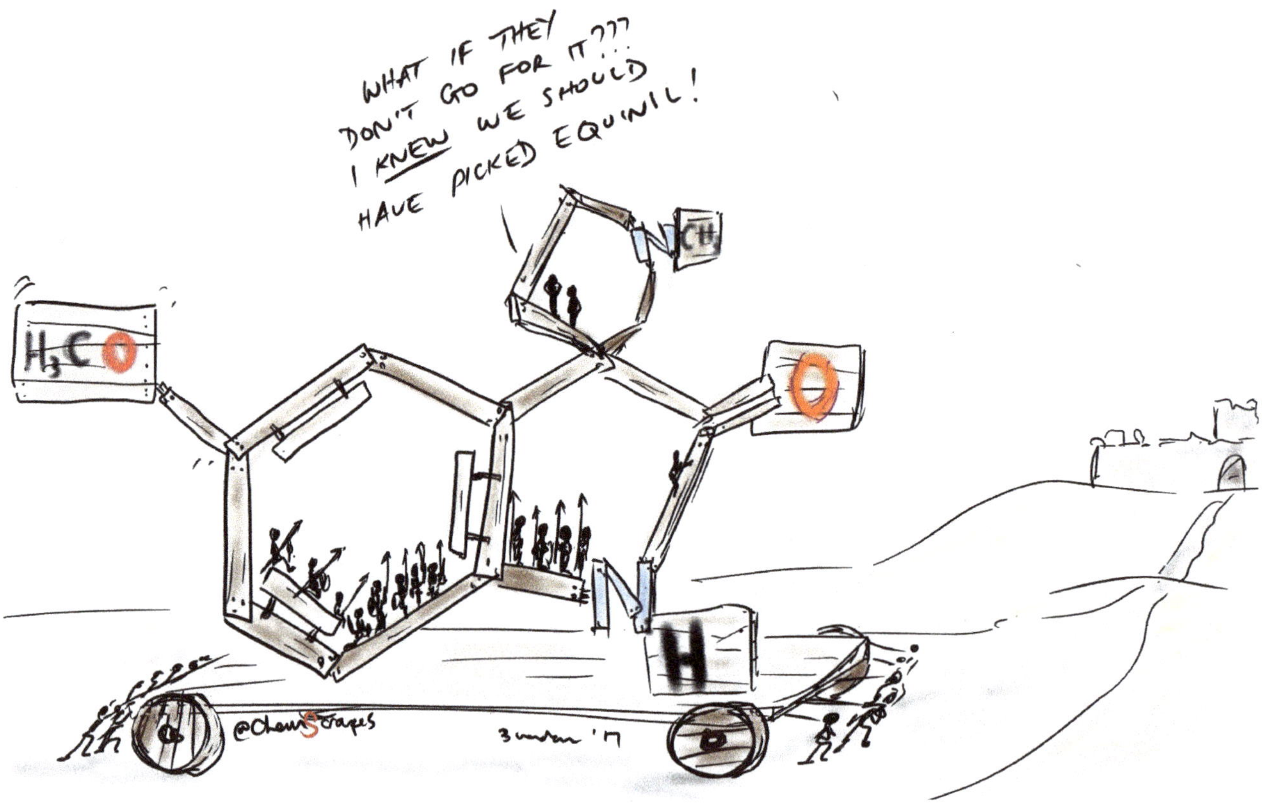

Trojan Horsfiline (2017)

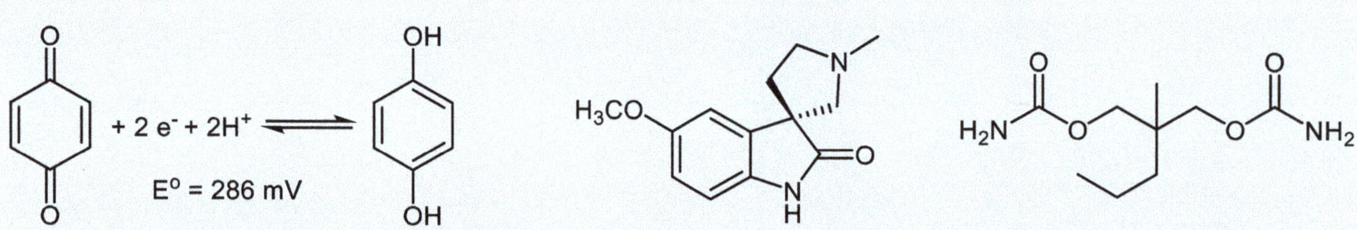

Quinone *Hydroquinone* *Horsfiline* *Equinil*

Spice girls (2017)

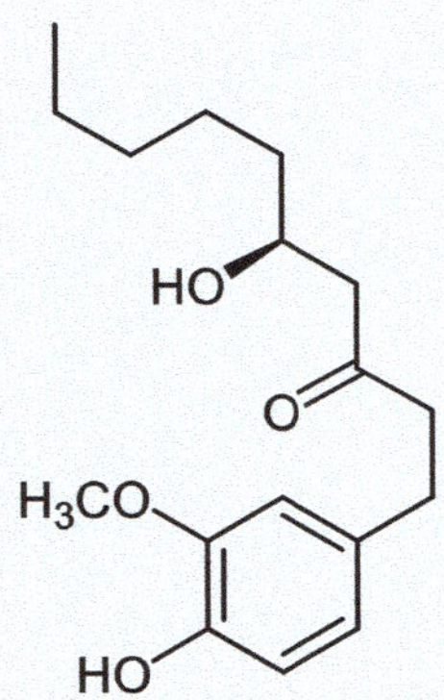

Gingerol
(from ginger)
Ginger Spice

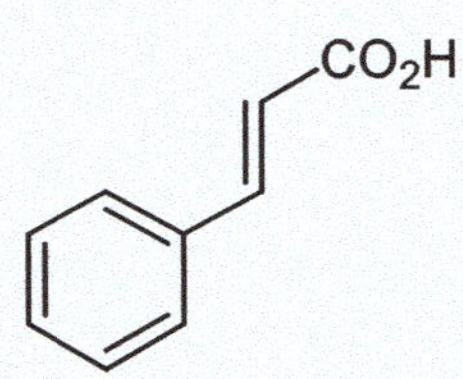

Cinnamic Acid
(from cinnamon)
Sporty Spice

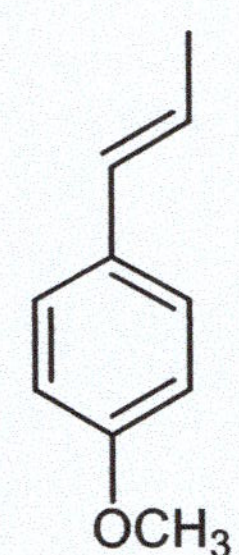

Anethole
(from star anise/fennel)
Posh Spice

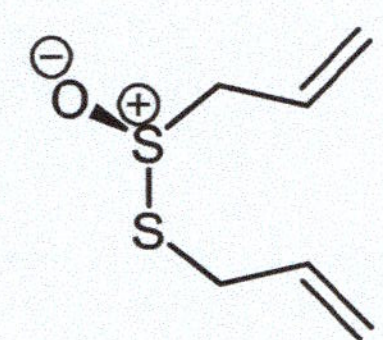

Allicin
(from garlic)
Baby Spice

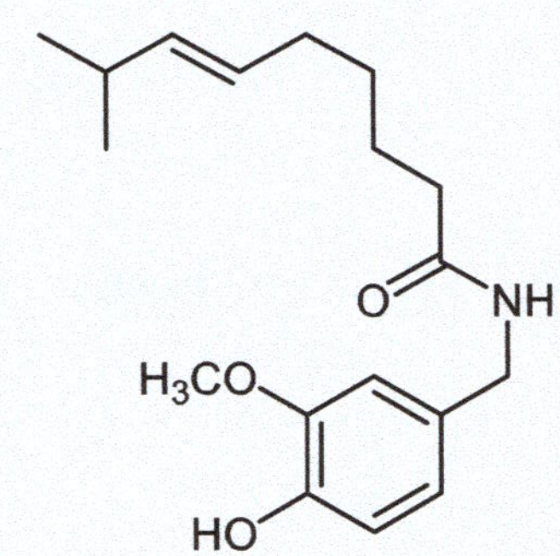

Capsaicin
(from chilli peppers)
Scary Spice

The Baeyer-Villiger People (2017)

The Baeyer-Villiger reaction is the overall insertion of an oxygen atom between the carbonyl of a ketone (Construction Worker) and the adjacent carbon atom to afford an ester (or in this case, a lactone - the GI). Oxidants such as MCPBA (m-chloroperbenzoic acid) or hydrogen peroxide (Chief and Policeman respectively) are commonly employed for this purpose. Over the years, the mechanism of the reaction has been heavily investigated, with a number of intermediates proposed and debated in the literature—the Cowboy is the perhydrate intermediate proposed by Dilthey and co-workers in the 1940s.[1] *Some may be wondering why N,N-dimethylaminopyridine (DMAP, Leather Man) makes an appearance as a band member because it is not typically used in Baeyer-Villiger reactions. The inspiration for this cartoon actually came when I was reading a paper on catalyst control over the selectivity of Baeyer-Villiger reactions, where DMAP was used generate peroxy acids from aspartate residues and hydrogen peroxide.*[2] *As soon as I saw this, Leather Man's nitrogenous crotch popped into my head, so DMAP is actually the founding member of The Baeyer-Villiger People.*

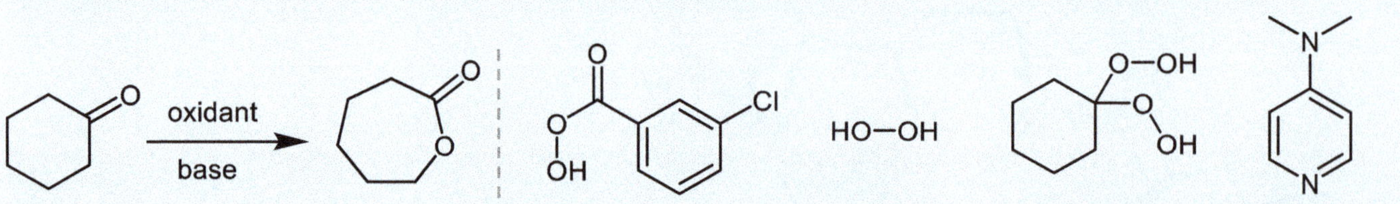

The Baeyer-Villiger Reaction | m-*chloroperbenzoic acid* | *Hydrogen peroxide* | *Perhydrate Intermediate* | N,N-*dimethyl aminopyridine (DMAP)*

1. *As read in* Renz, M.; Meunier, B. 100 Years of Baeyer-Villiger Oxidations. *Eur. J. Org. Chem.* **1999**, 737-750.
2. Romney, D. K.; Colvin, S. M.; Miller, S. J. Catalyst Control over Regio- and Enantioselectivity in Baeyer-Villiger Oxidations of Functionalized Ketones. *J. Amer. Chem. Soc.* **2014**, *136*, 14019-14022.

The curse of lactic acid (2017)

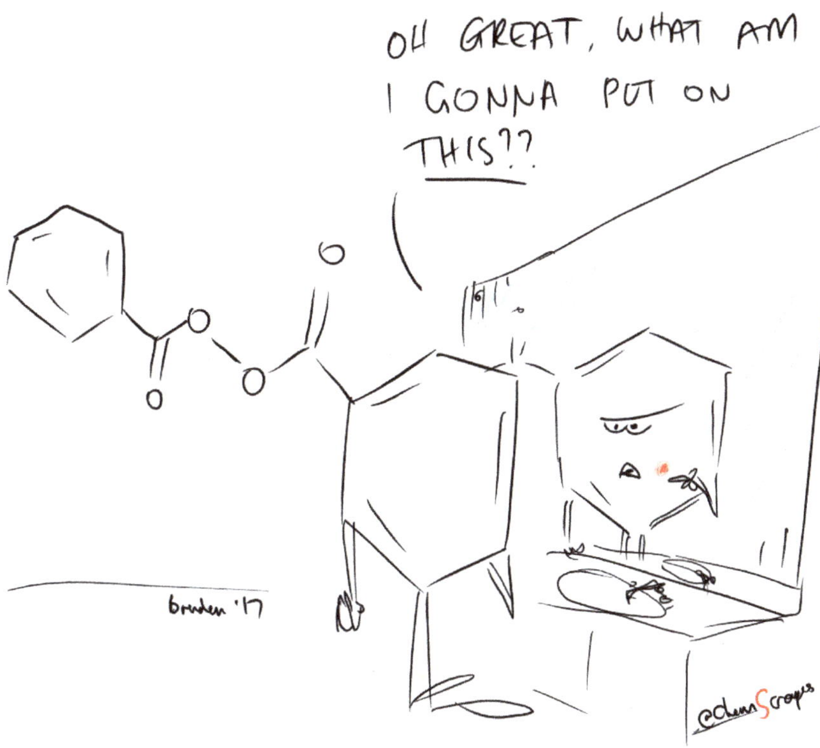

The curse of benzoyl peroxide (2017)

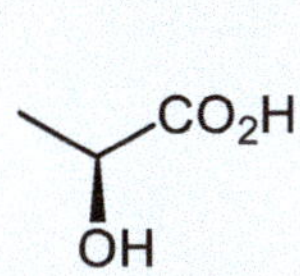

Lactic acid

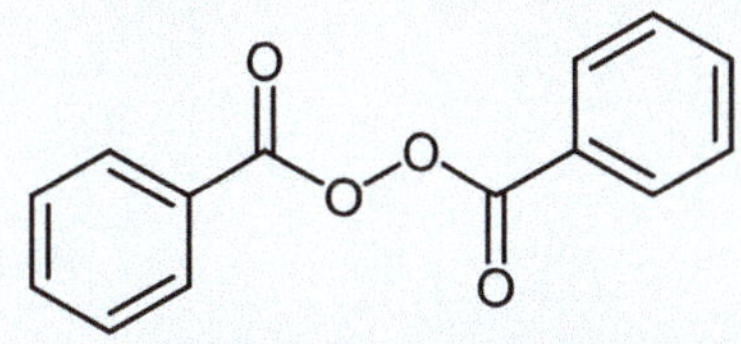

Benzoyl peroxide

Stapled Peptides: the punk rockers of the peptides world (2017)

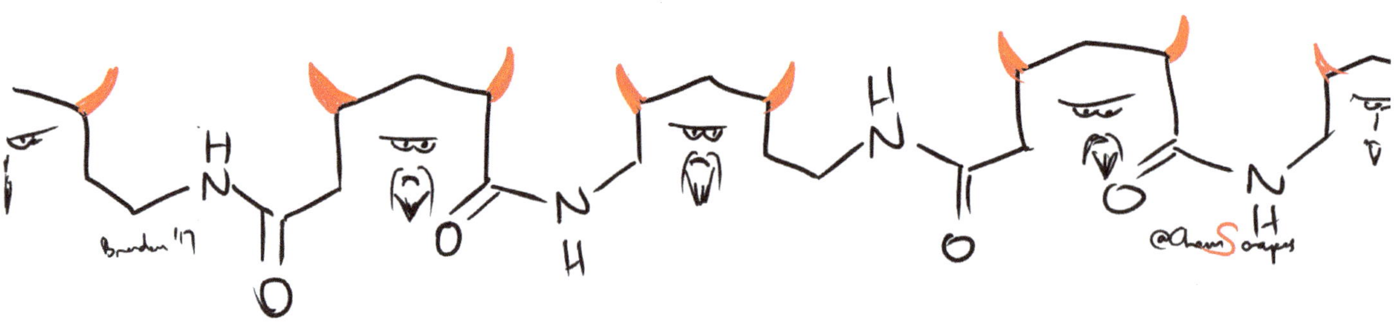

Nylon-666 (2017)

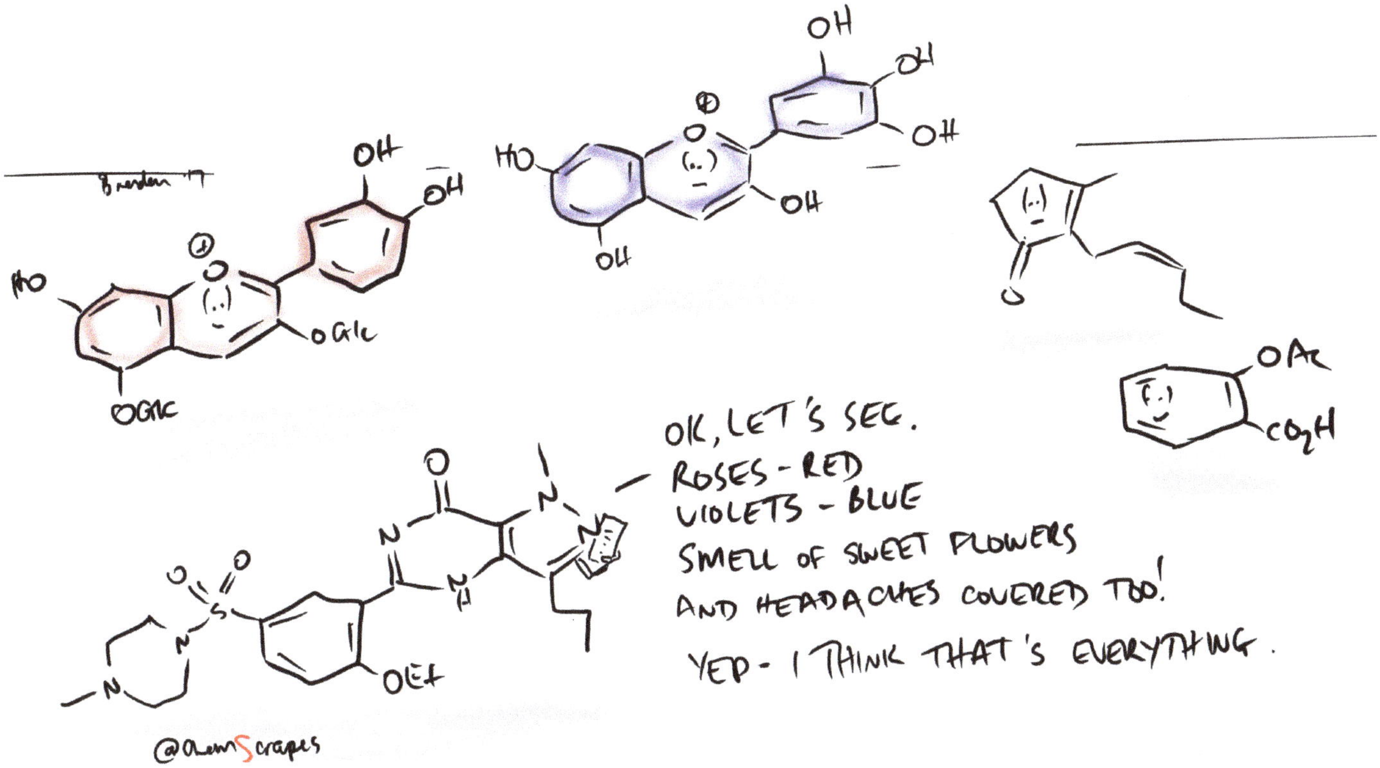

Viagra valentine (2017)

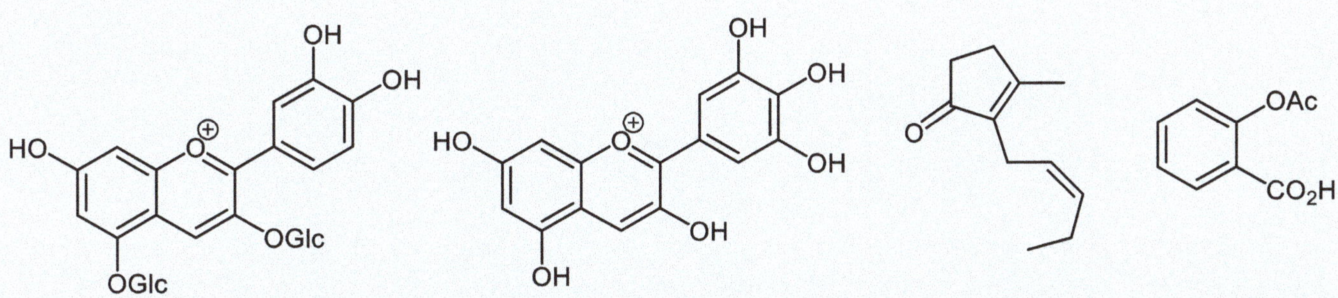

Cyanin *Delphinidin* *Jasmone* *Aspirin*

Viagra has just one thing on his mind for Valentine's day, and wants to be sure he has all possibilities covered to impress his love. From left to right, the colour of roses (cyanin—Glc is glucose), violets as well (delphinidin), the smell of sweet jasmine (jasmone) and finally, aspirin—just in case she has a headache.

Triphenyl phosphine (TPP) (2017)

In response to US withdrawing from Trans Pacific Partnership soon after Trump took office.

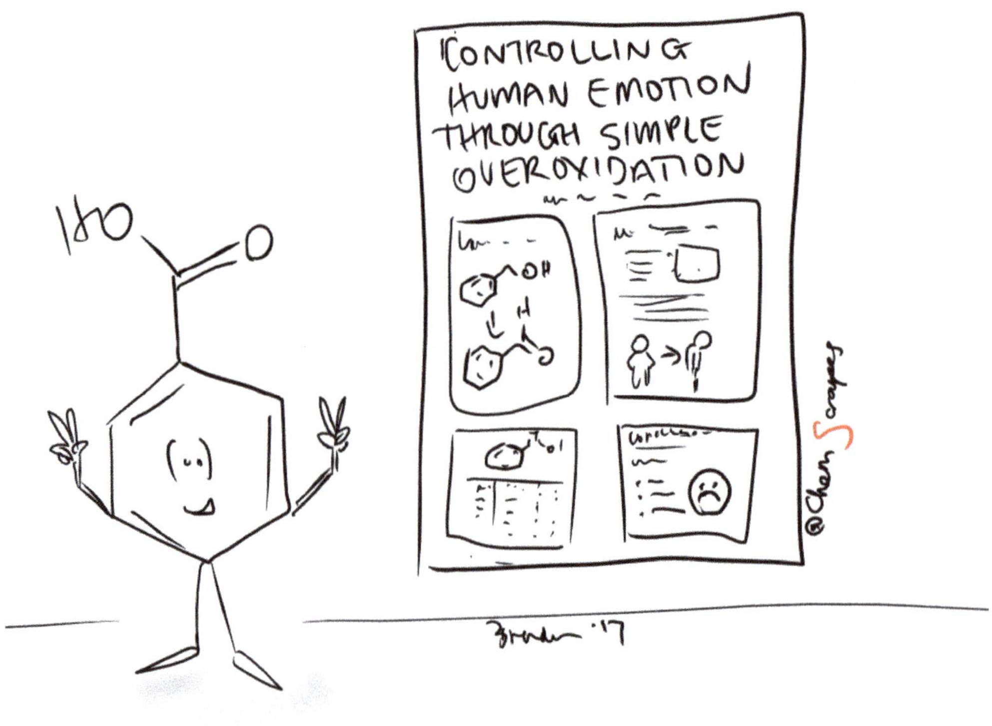

At the "molecular basis of behaviour" conference (2017)

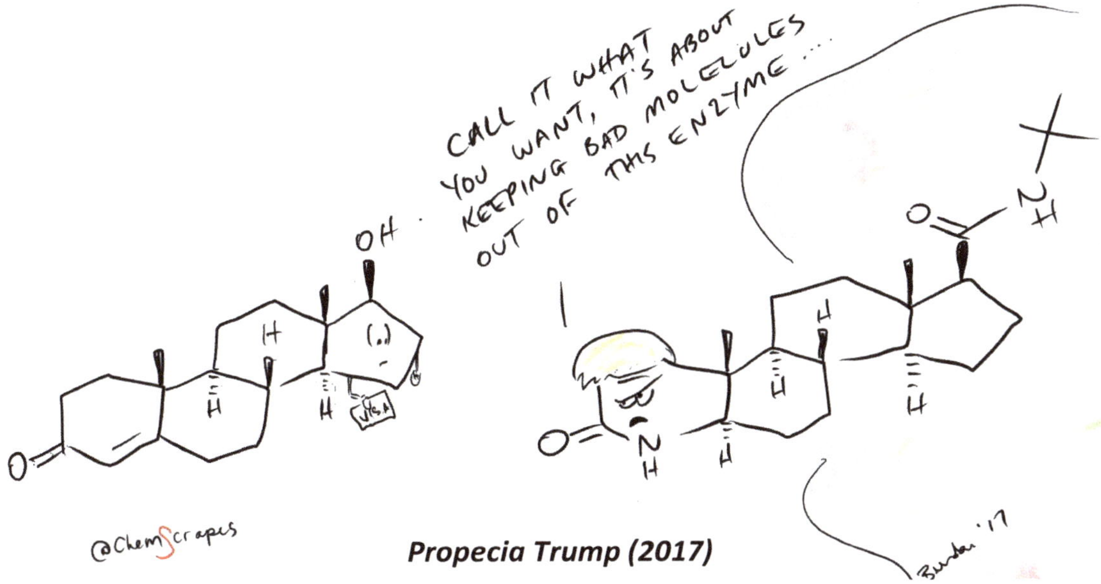

Propecia Trump (2017)

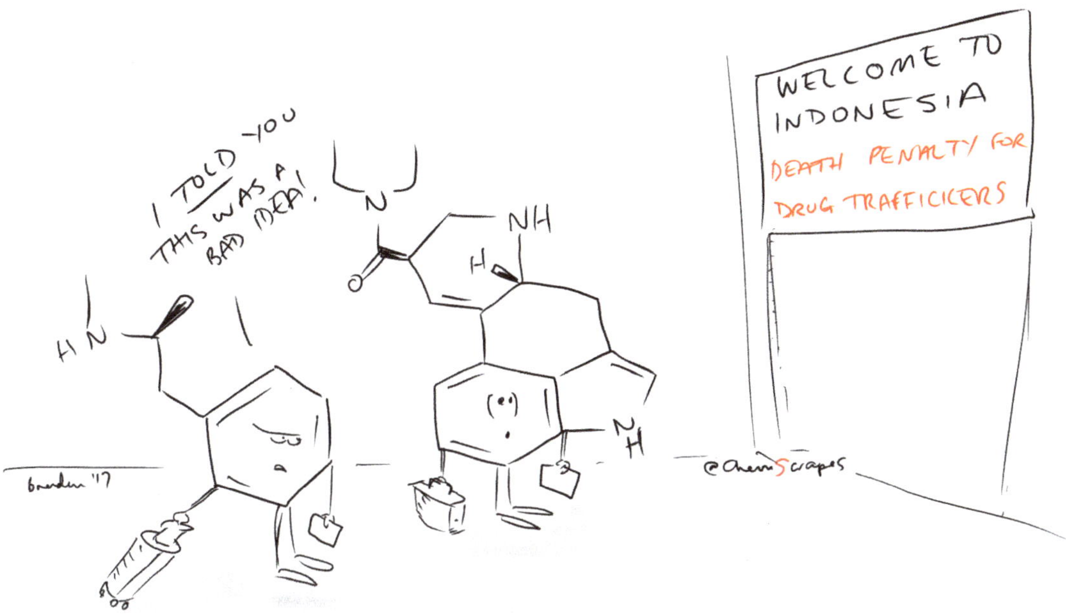

The curse of alkaloids (2017)

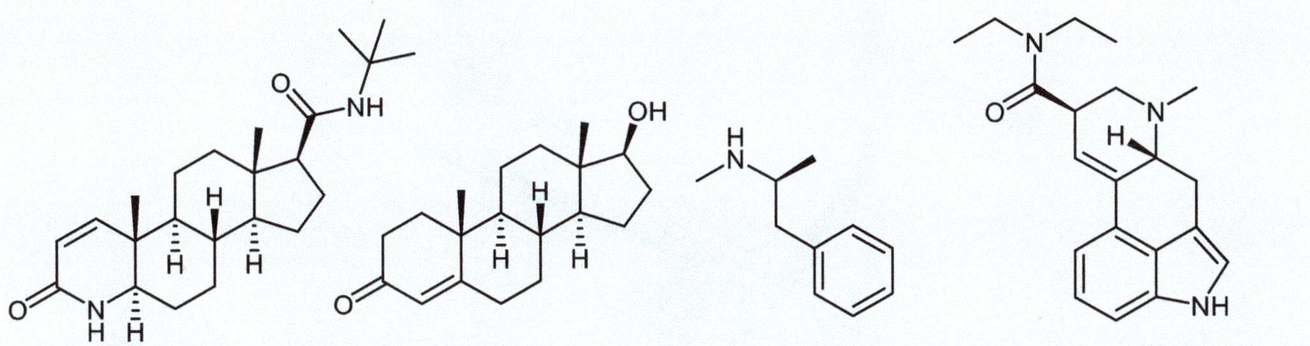

Propecia *Testosterone* *Methamphetamine* *Lysergic acid diethylamide (LSD)*

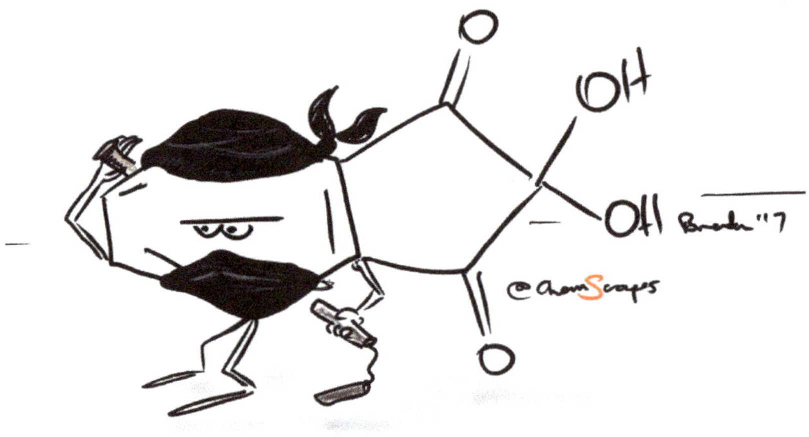

Ninjahydrin (2017)

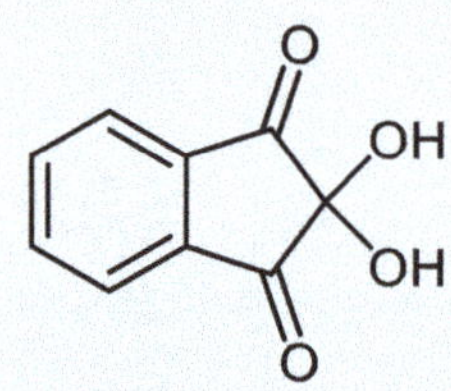

Ninhydrin

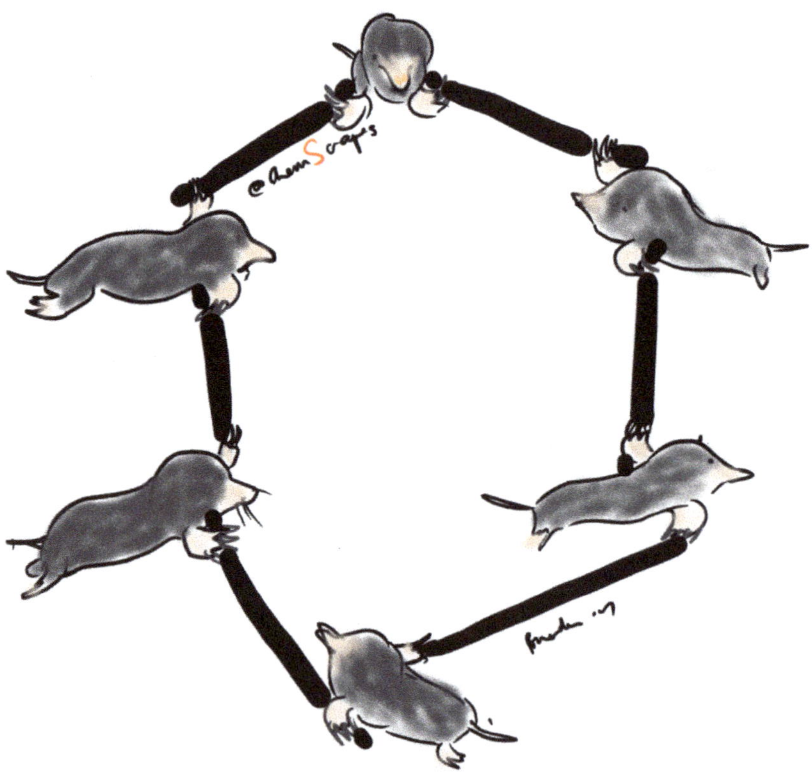

A mole-ecule (2017)

Drawn for Mole day (October 23rd)

Tryptophan float at the Macy's Thanksgiving Parade (2017)

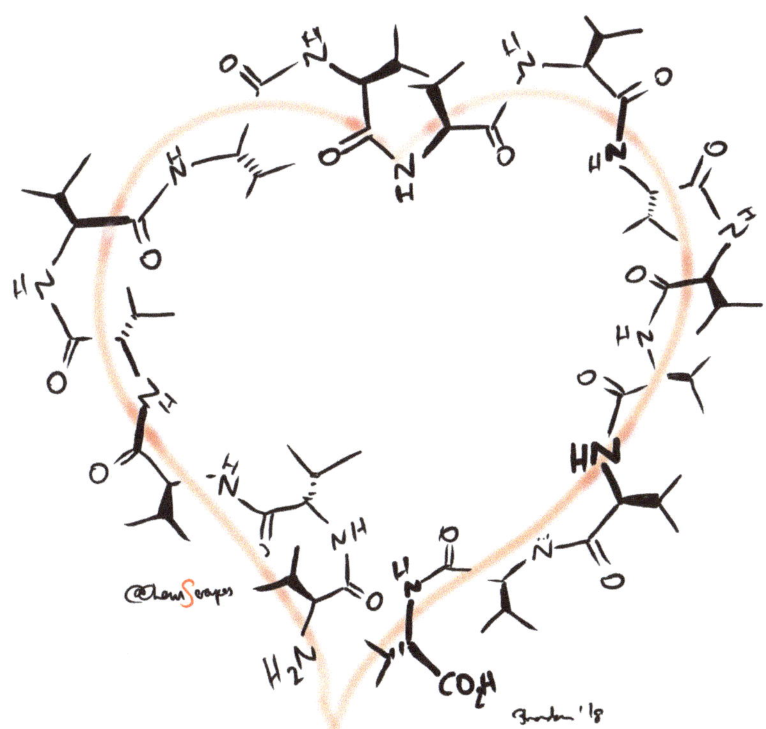

Val-entwines day (2018)

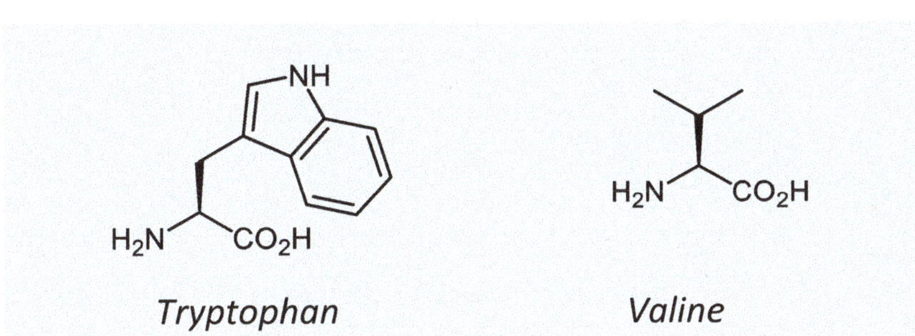

Trouble brewing on the road to Bethlehem (2017)

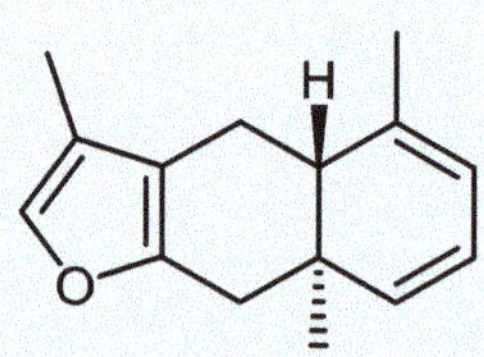

Furanoeudesma-1,3-diene

(from myrrh resin)

79

Au

Gold

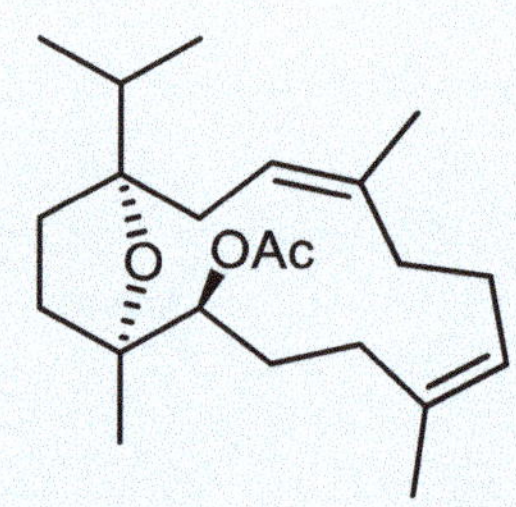

Incensole acetate

(from frankincesne resin)

Thionitrite's watch (2018)

Other Things Chemistry

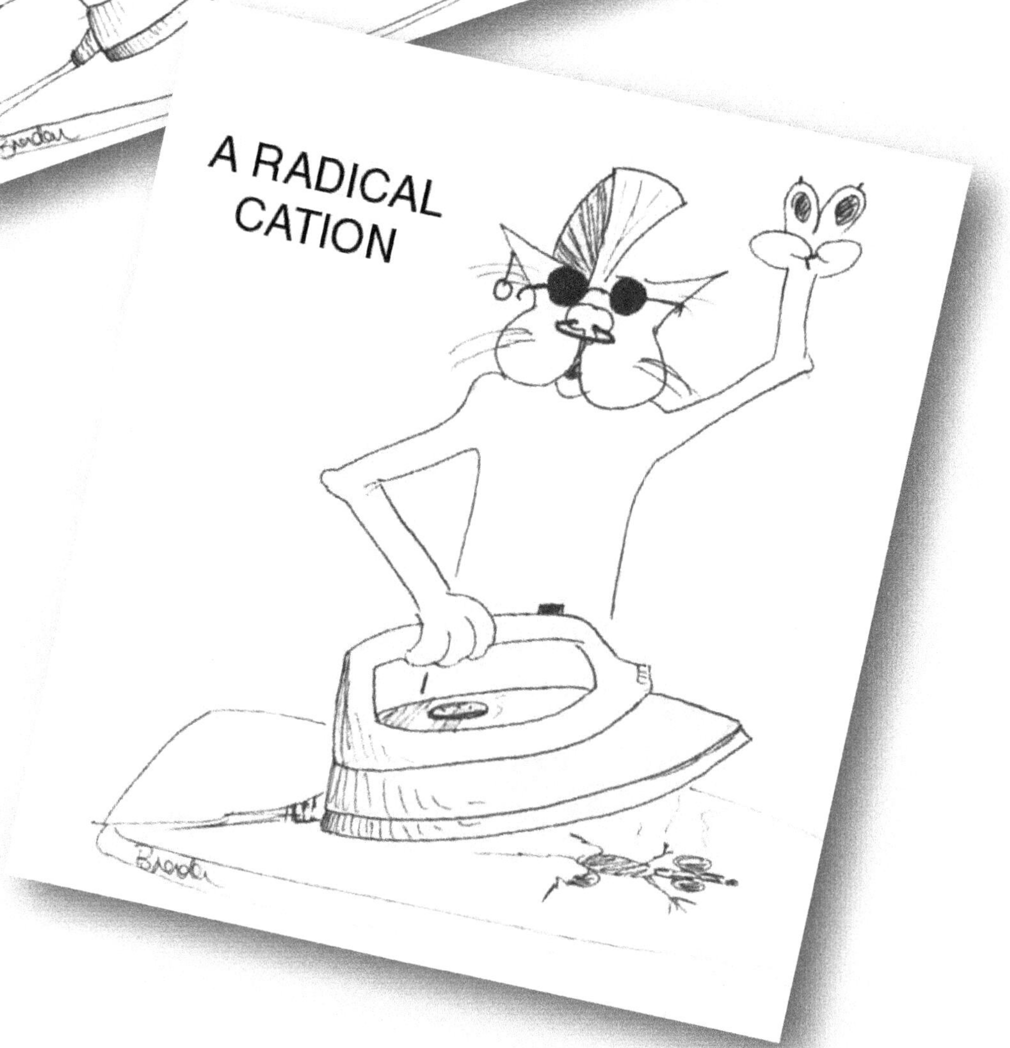

Cations (1999)

Marc Antimony and Chloropatra (2001)

Solid phase extraction (2001)

Scavenger Resins (2001)

Scavenger Resins (2018)

Ion exchange resins (2001)

Cross, linked, polymer beads (2003)

"Yes, that's them. The little doublets on the right."

Proton maternity wards (2003)

*"**ACCELLERATE** you idiot!"*

At the cyclotron (2003)

Chem abs (2004)

SOMO the Kid, scourge of the wild frontier (2004)

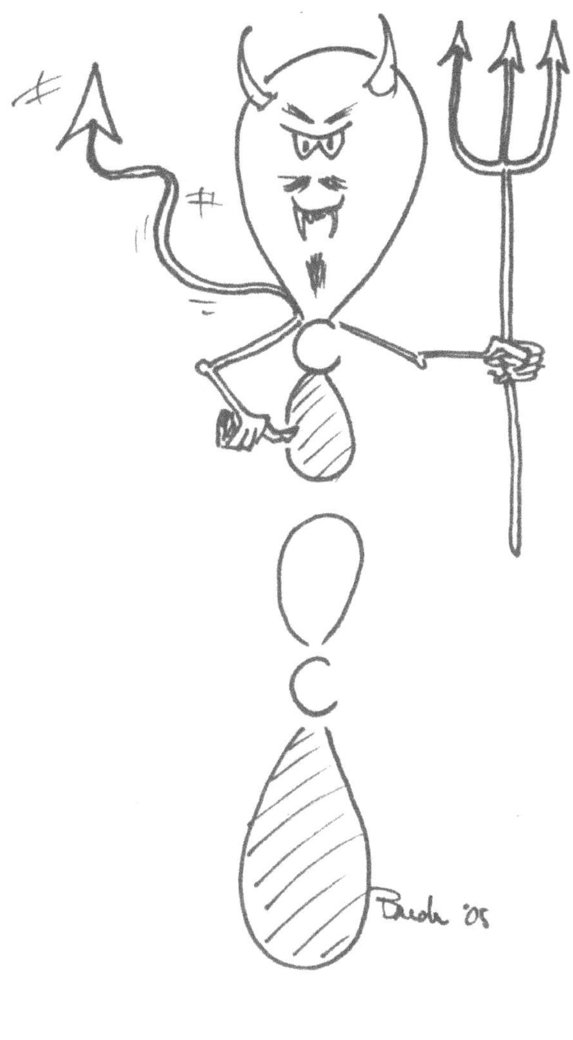

The antibond (2005)

Single crystal bars (2003, 2017)

*Originally drawn in 2003 (above left), the cartoon entitled "Single Crystal Bars" was revamped for consideration as a **Sketch Chemistry** panel (above right). Following submission, a few different ideas were investigated in order to develop more of a relationship between the single crystals and maybe slip a few puns in. Right-top shows a more cheesy pick-up line being used and the female crystal appearing to be a bit bothered by the whole situation. Right-bottom shifts the focus to the female crystal, disconcerted by the early arrival of her blind date. Below is a rough sketch which completely removes the male crystal from the picture, but in doing so introduces a fairly unflattering stereotype. We made the decision that perhaps this cartoon was not entirely suitable for **Sketch Chemistry** and the idea was appropriately dropped.*

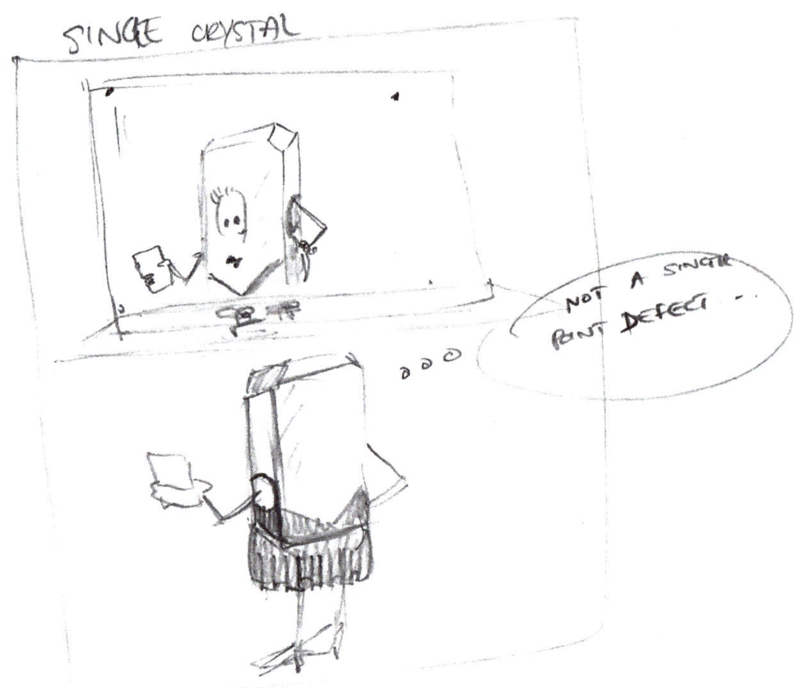

Single crystal bars 2 (2018)

Single crystal life (2018)

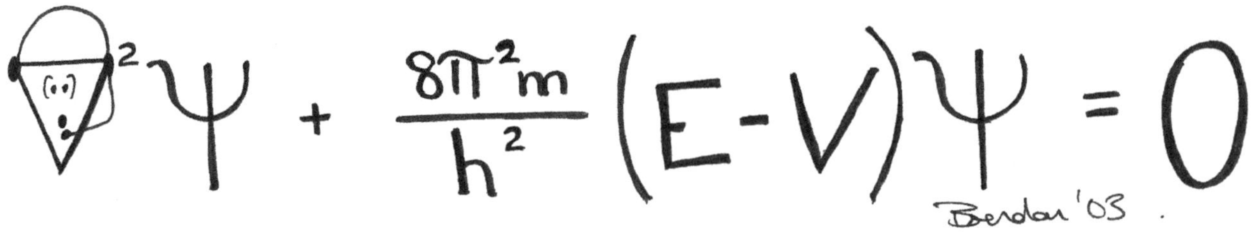

Operator (2003)

Micelle Jackson (2003)

Micelle Jackson (2016)

Left: ***Nanna Particles (2013)***

Below: ***Nanna particles (2018)***

Angry allotropes (2012)

Indifferential equations (2013)

"I'm sorry, it's just not gonna work. You are becoming SO negative……there is just no spontaneity any more!"

Thermodynamics (2013)

Doesn't matter (2016)

Sour Grapes (2016)

Chem Brexit (2016)

Voted as one of the top 10 Chemistry Twitter moments of 2016 by Chemical and Engineering News.

Chem Article 50 (2017)

Vote to leave the EU:

For: 51.9%

Against: 48.1%

Natural abundance of isotopes:

^{79}Br: 50.9%

^{81}Br: 49.1%

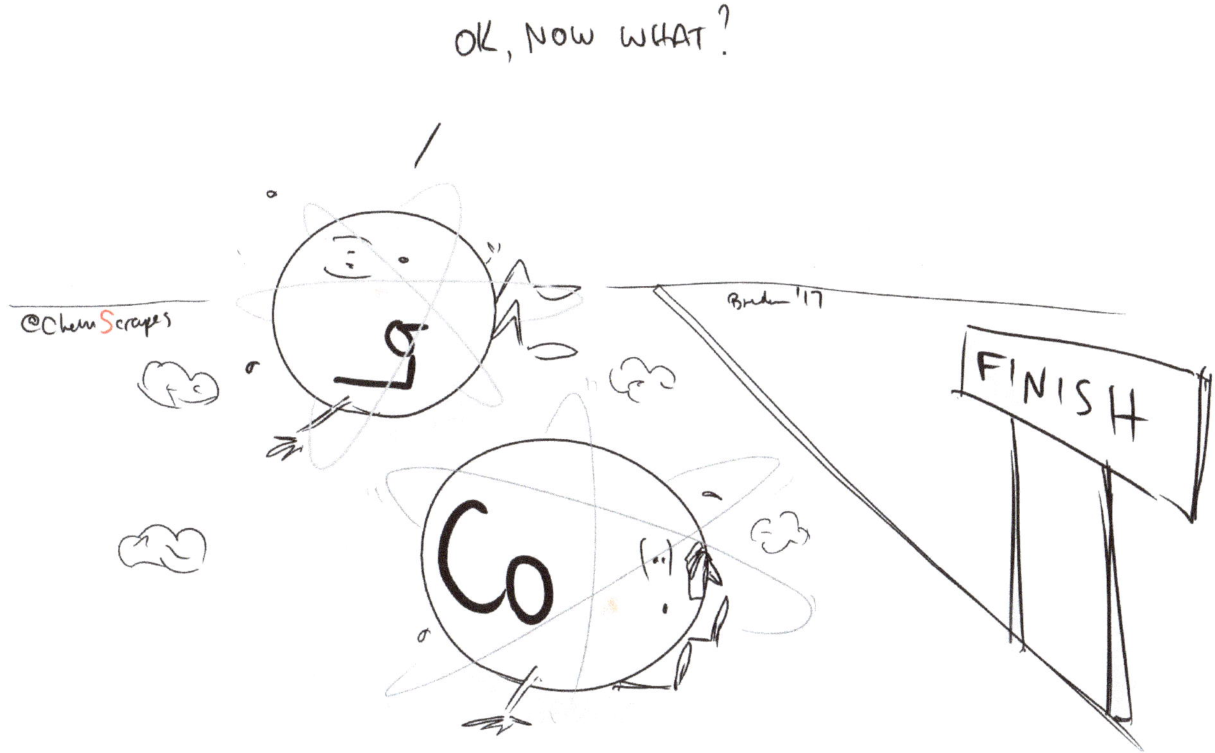

Chem Elections (2017)

In response to the close result between Labour and Conservatives in the 2017 UK elections.

Chem G20 (2017)

On the occasion of President Trump pulling out of the Paris accord.

Quadrupole (2016)

The frank, condom principal (2016)

This is a play on the "Franck-Condon Principle" in spectroscopy which deals with vibrational energy changes between electronic states of molecules.

ChemScrapes emoticons (2016)

ChemScrapes chess set (2017)

Atomic card games (2017)

Surf act ant (2017)

MUFC's atomic fanclub (2017)

Liverpool's atomic fanclub (2017)

Atomic U2 (2017)

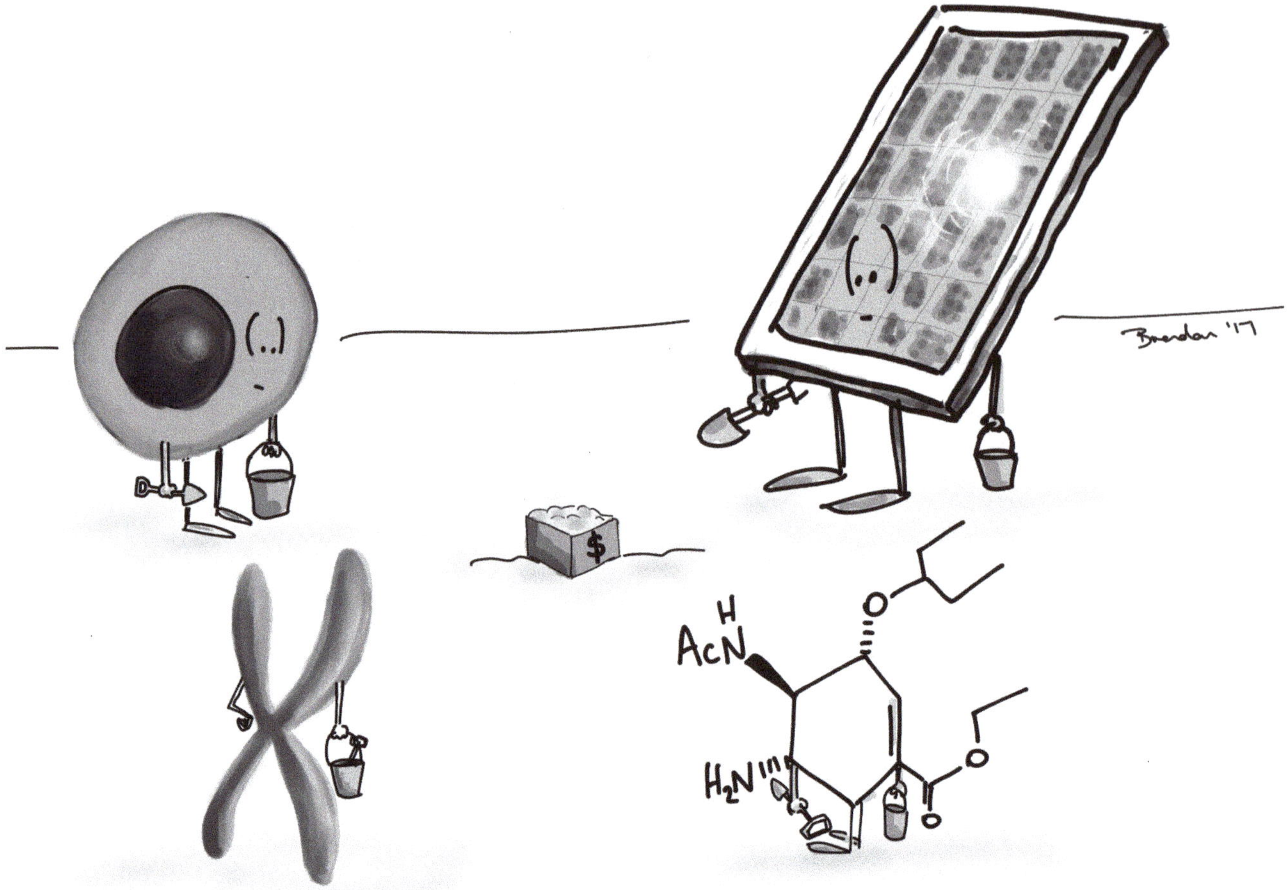

Science under Trump (2017)

In response to Chemical and Engineering News article on the reducing size of the funding sandpit under Trump.

Atomic Chinese New Year (2018)

Glass and Lab

Concentrating HCl (2001)

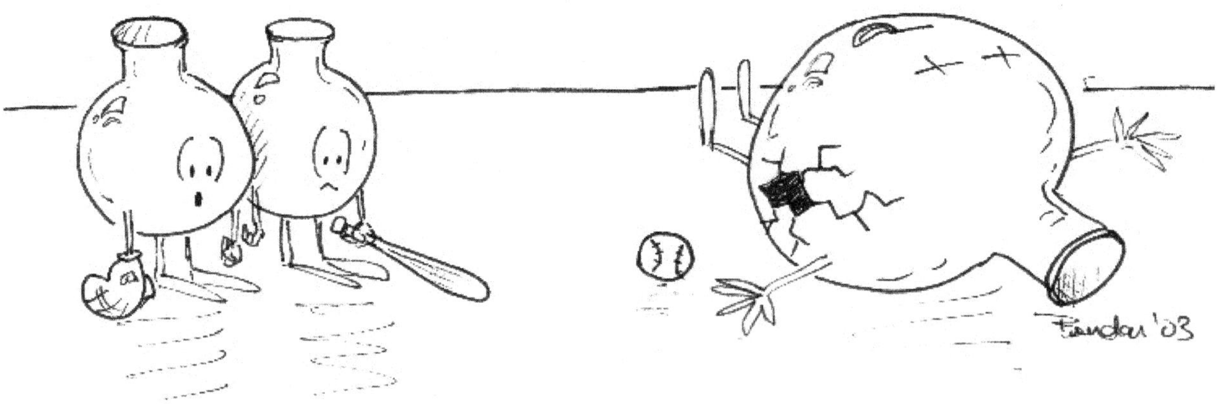

"UH-OH!"

Oops (2003)

"I'm telling you, you're NUTS! They'll never go for it!"

Try hard (2003)

Glassware amusement parks (2003, augmented 2016)

The other HPLC columns were both shocked and disapproving at Pamela's silica implants.

Silica implants (2003)

Acid test (2003, recoloured 2017)

At the glassware erotic film festival (2003)

At the glassware erotic film festival (2016)

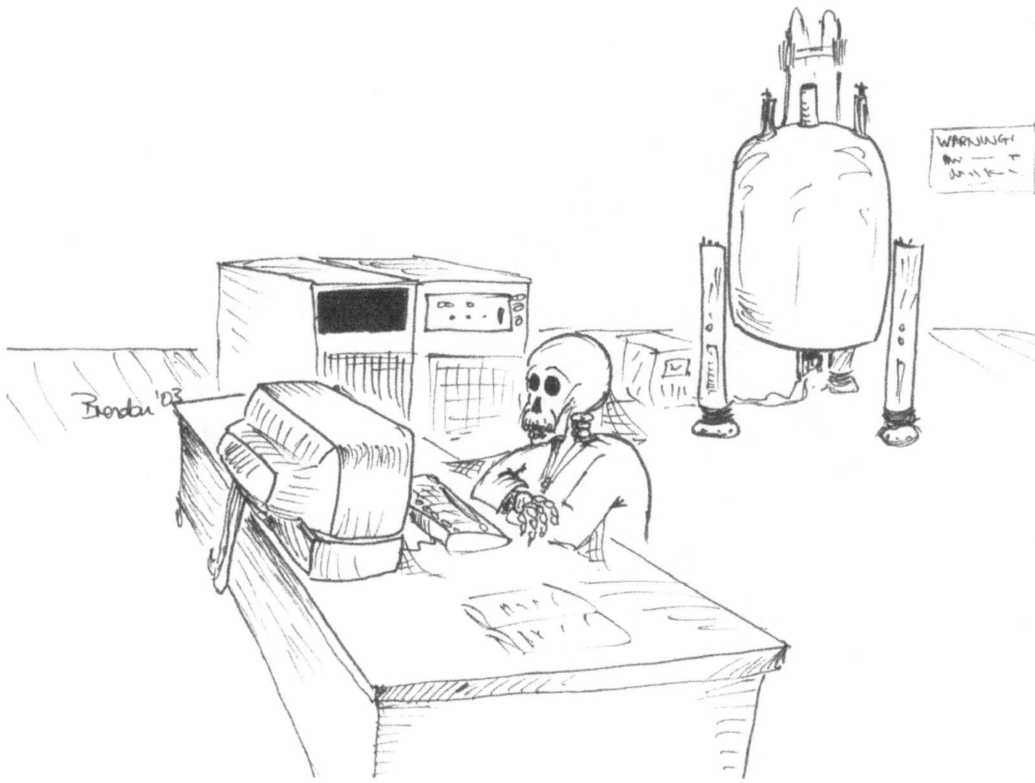

Autoshims (2003)

"I TOLD you not to play so rough. Now sweep up your brother an put him in the recycle bin"

Rough play (2005)

Dirty Glassware (2006)

***Sprung** (2010)*

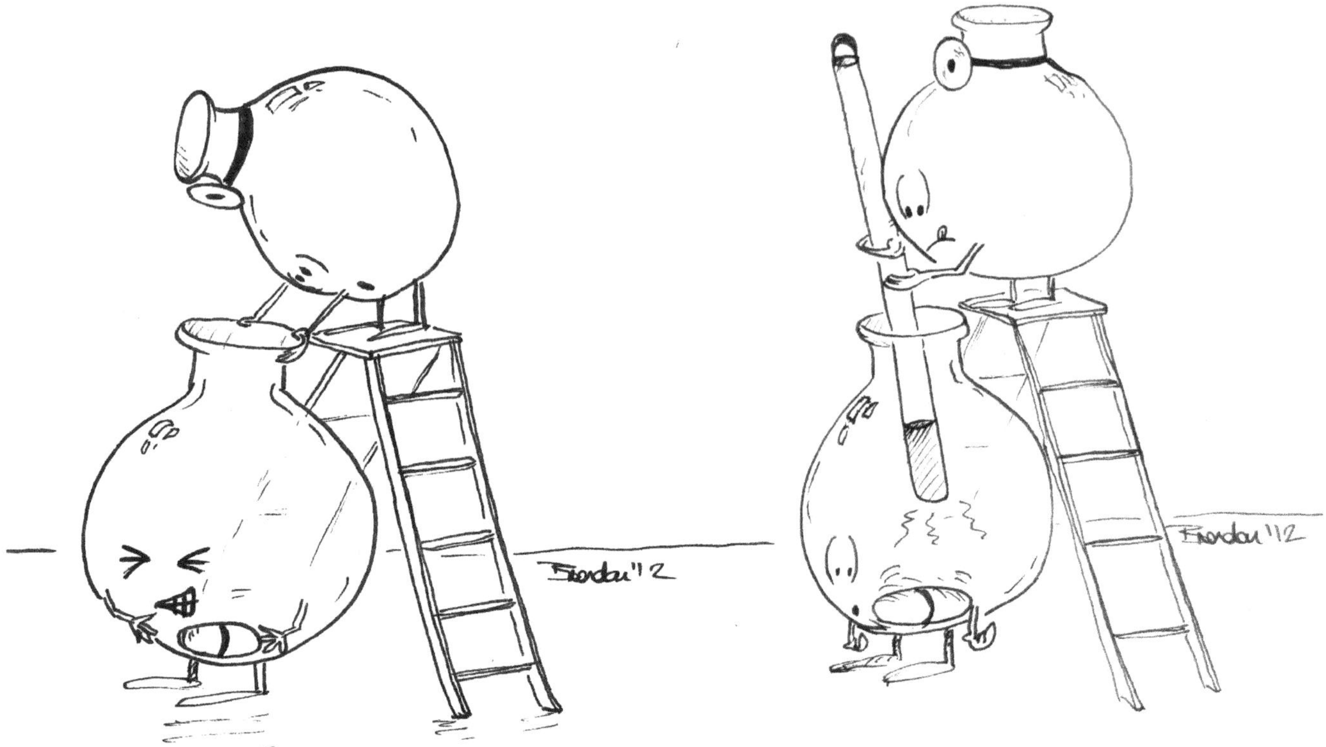

Flask GPs (2012)

"I was going to wait to tell you, but I've already started to show."

Good show (2012)

"Hey Fred—your crack's showing!"

Crackup (2012)

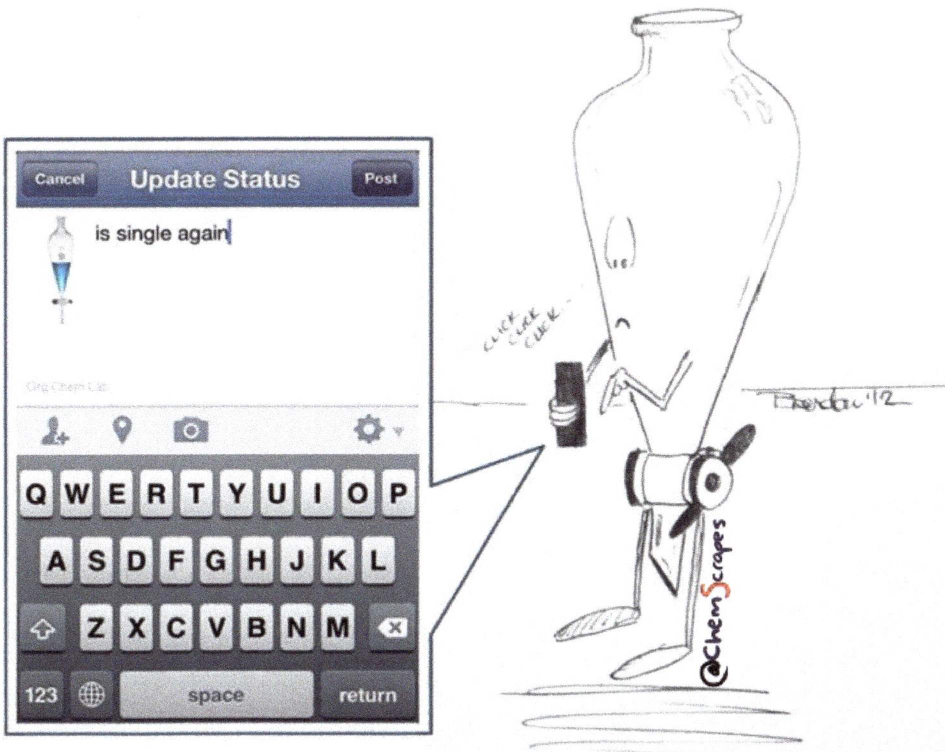

Curse of the separating funnel (2012)

"Your "laminar mixing" ain't welcome 'round these parts! If you have any sense, you'll turn around and haul that square arse of yours out of here."

Not going with the flow (2012)

Not going with the flow (2018)

Busting (2015)

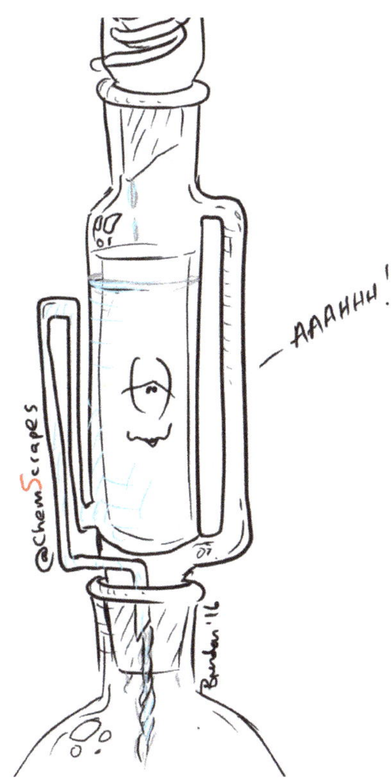

Soxhlet (2016)

Stuck (2016)

Christmas cheer (2017)

People

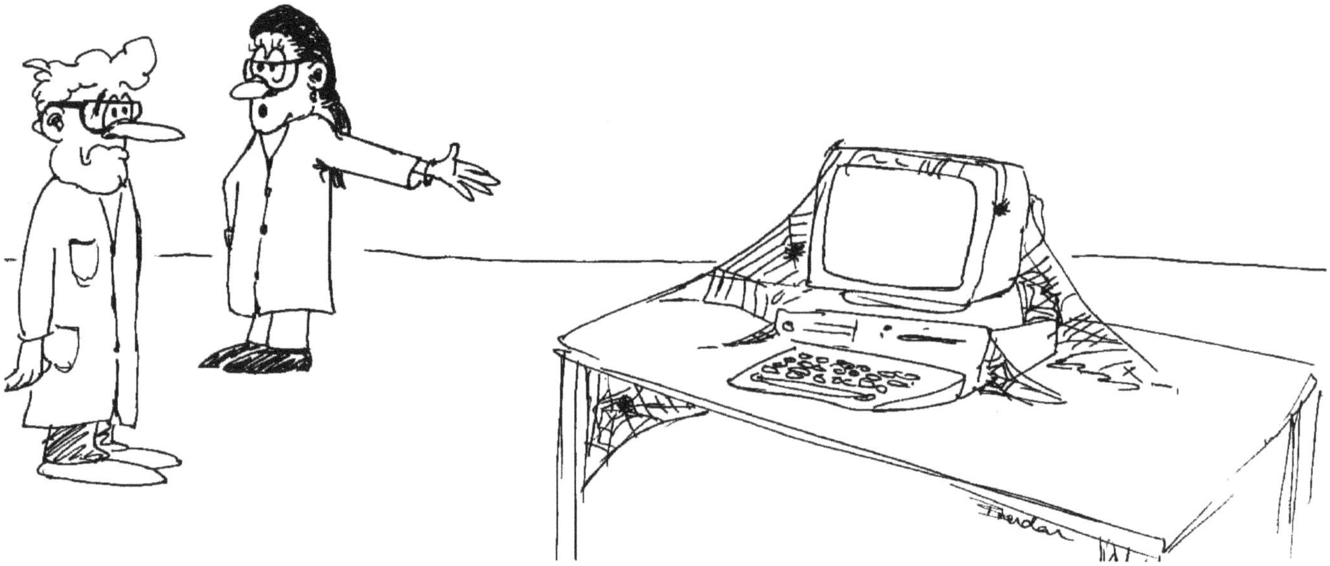

"So, you will be in charge of putting the inventory on the web...of course it could be argued the job is already half done."

Web (2001)

The other chemists were insanely jealous of Anita's one pot procedure.

One pot (2001)

Shipwrecked (2001)

Shipwrecked (2003)

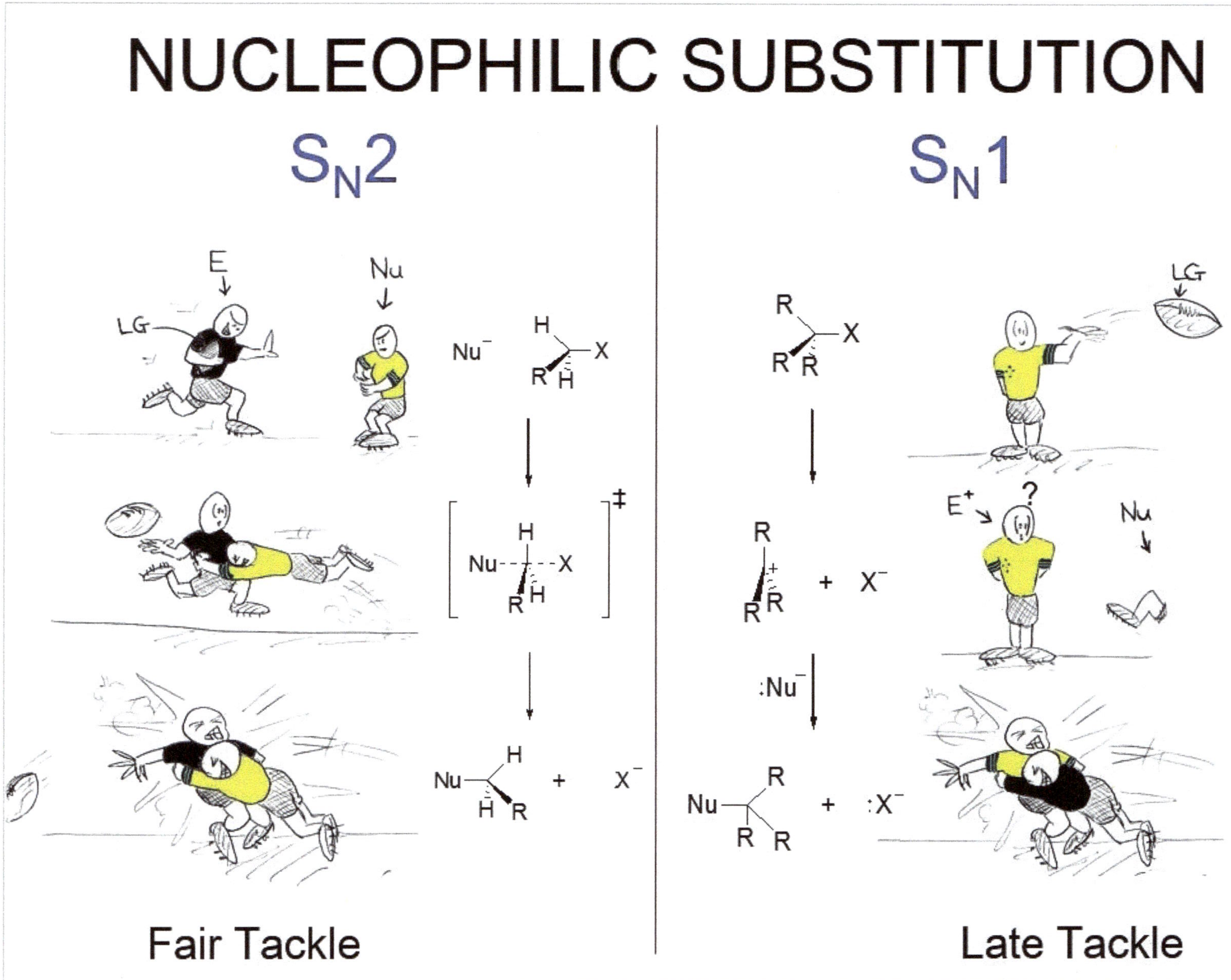

From teaching: Ruganisms (2004)

After moving downunder from North America, it took just one embarrassing mistake for Tim to hate the abbreviation "NMR".

Enema (2004)

"Oh Great! Now what am I going to tell the kids?"

Santa's hood (2010)

"I've gone through the scenarios you suggested, identified all the potential hazards and put them into a risk assessment matrix. Carpet burn, STDs, back injury.......not to mention a significant choking hazard! Look, unless you can get this down to medium there's definitely no action tonight."

Risky Business (2012)

"Monthly report? No! This is just the risk assessment for the first experiment."

Paperwork (2013)

Significant moments in the history of chemistry: 1885 - Raphael Dubois coins the names Luciferase and Luciferin to describe the enzyme/molecule system responsible for bioluminescence observed in the South American Firefly, and becomes the first trademark infringement case in chemistry.

"Correct! My client is suing you for unauthorised use of his name, which he had registered as a trademark in the time of King James."

Devil in the details (2016)

Hell's Laboratory (2016)

Eu d'toilet (2016)

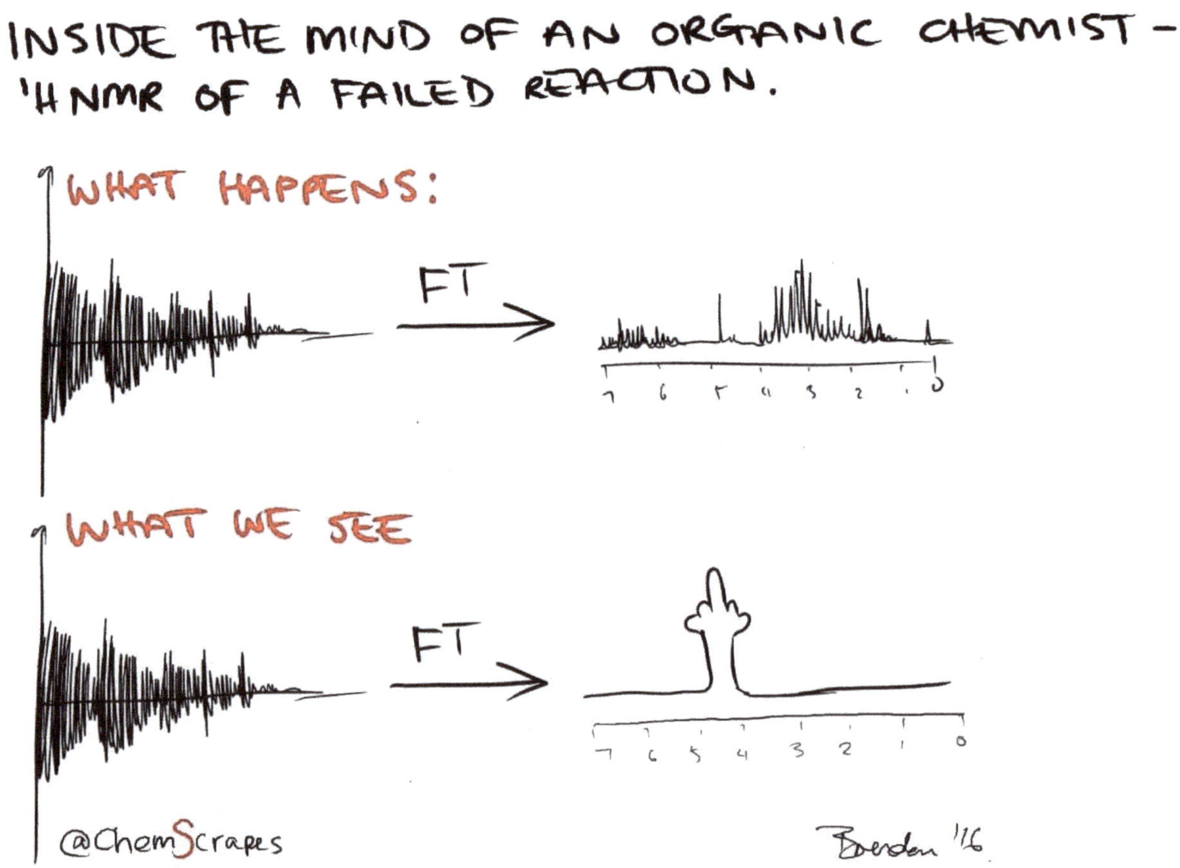

Effed-T NMR (2016)

Academia: Then and Now

Papertrails (2017)

Academic trick or treat (2017)

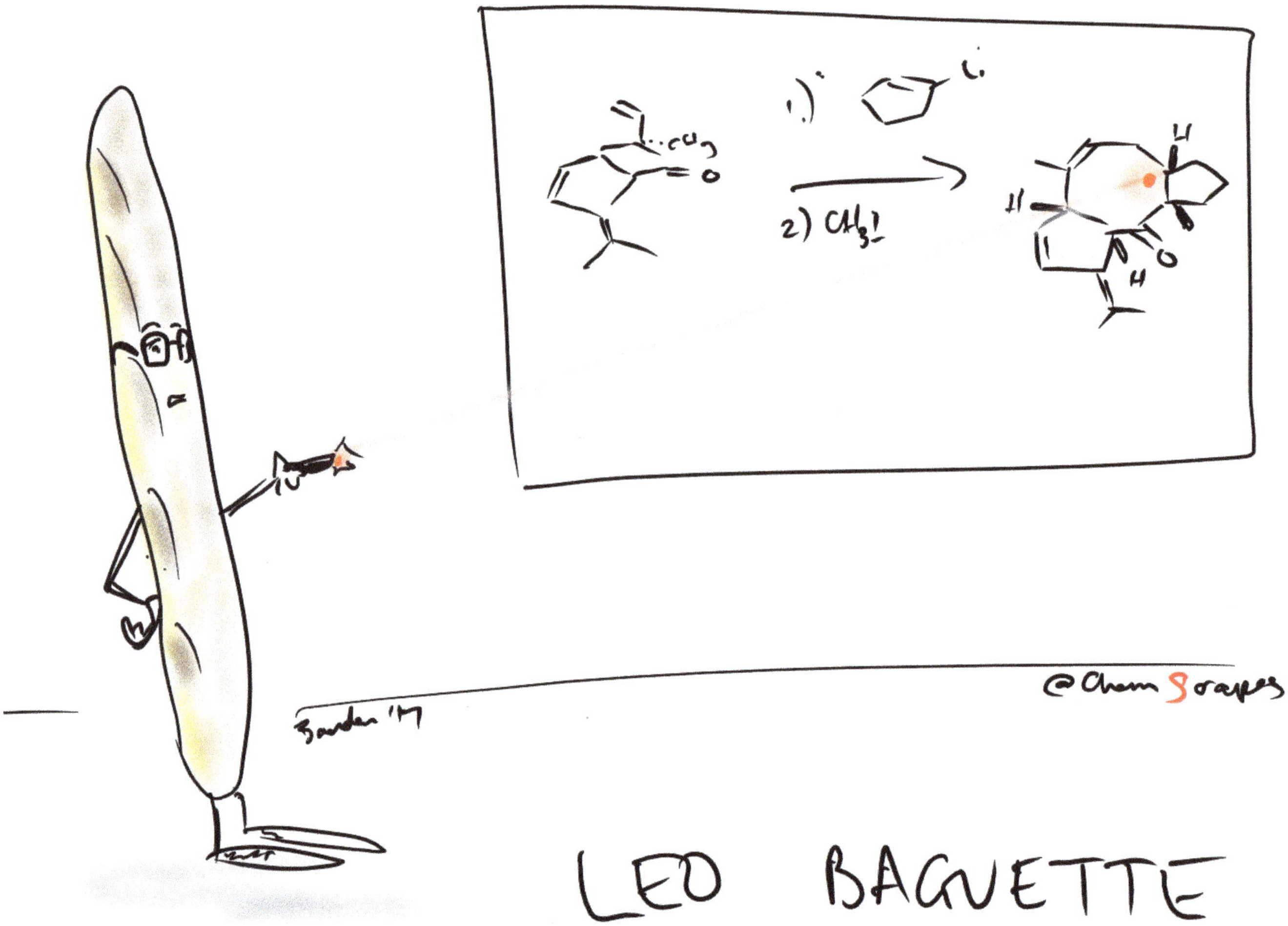

Leo Baguette (2017)

A pain-ful pun on Leo Paquette (Ohio State University)

General Science

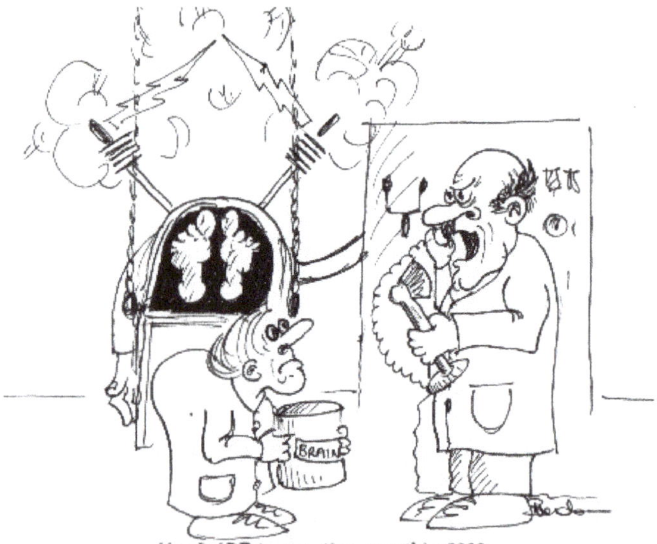

You DARE to question my ethics????
Of COURSE I'm not involved in any human cloning!!

Left: Ethics (2000)

Below: Ethics (2018)

"You DARE to question my ethics? Of course I am not involved in human cloning experiments!"

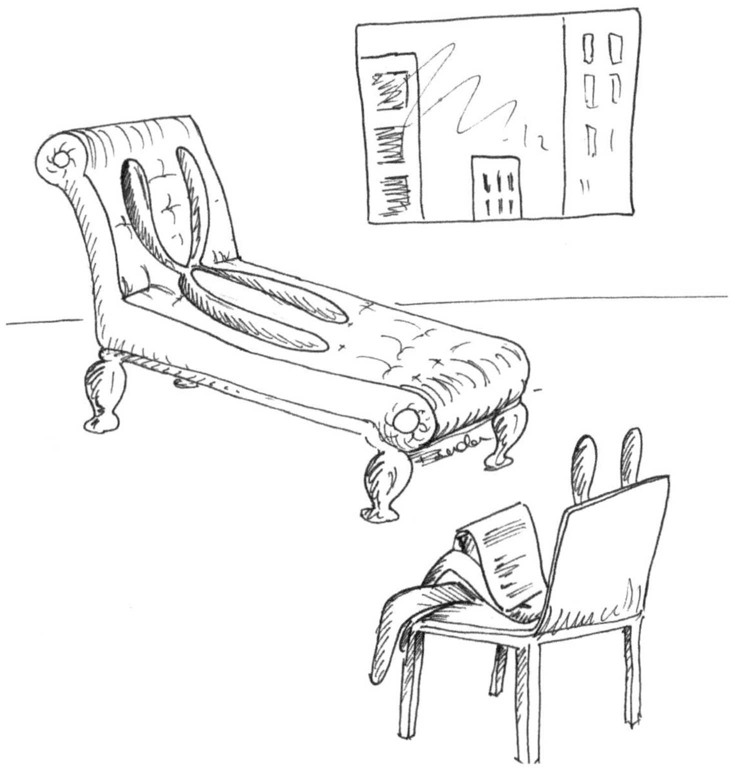

"So, you say it all began when you separated from your daughter cell...."

Gene therapy (2001)

Newton shows that for every action, there is an opposite, but not necessarily equal reaction.

Newton's rules (2003)

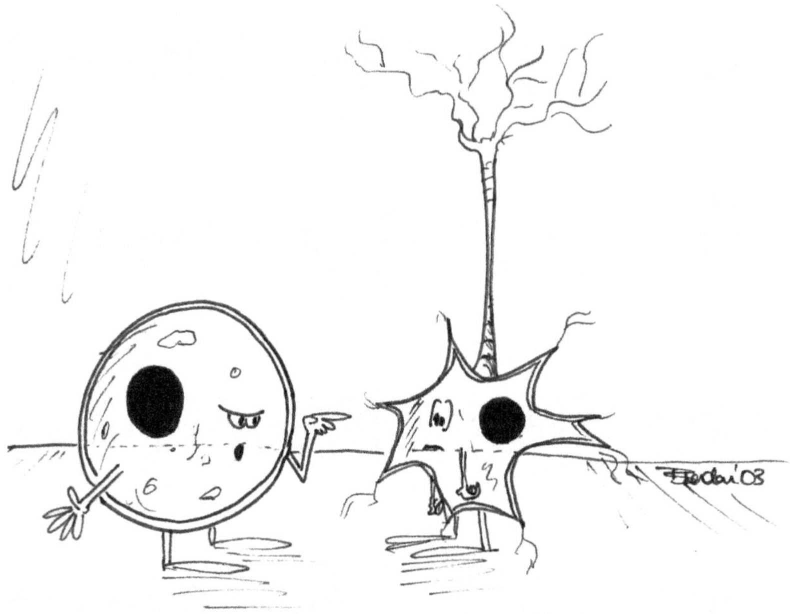

"You gotta lot of nerve showing up round here!"

Nerve! (2003)

Apoptosis: Programmed cell death (2003)

"You idiot Henderson! Human GENOME project....GENOME.....G-E-N-O-M-E!"

Human g-nome project (2004)

The gene pool (2004)

"I thought you were meant to be a carbon SINK!"

Kyoto unravelled (2006)

(Caption provided by Professor Paul Teasdale-Spittle, Victoria University of Wellington)

"Yep! You can see it right there! Clear signs of fowling."

Marine biofowling (2012)

Genetic disorder (2013)

*This cartoon was drawn in response to a call from Elise Andrew (of I F**king Love Science (IFLS) fame) to insult her in the most imaginative way possible because she was sick of the bad language and threats of mutilation she received in response to her involvements in public debates regarding Creationism as a scientific theory.*

This scenario sees a fictitious institute (named after Richard Dawkins) being marched upon by a bunch of chromosomes carrying slogans against the theory of evolution. The slogans in the background on the left read "you will burn" and "IFLS bitch". The three most prominent chromosomes are XXY (Klinefelter's syndrome) which led to the cartoon's title - "Genetic Disorder".

As far as I know Elise Andrew has never seen this cartoon despite multiple posts on the IFLS Facebook page, multiple hashtag combinations/mentions on Twitter and more than a couple of direct e-mails over the years.

Drosophila MELONogaster (2017)

Soya Sorts (2018)

Soya "milk" by definition cannot be a milk as it is not produced by mammary glands. However, the industry continues to call it milk.

ChemScrapery

The word "ChemScrapery" essentially refers to the generation of cartoons in real time interaction with some of the online communities in Social Media. The art of ChemScrapery is anything that takes ten minutes or less to turn around, including any post drawing editing, poetry or captions. What follows are some of the fun moments from Twitter that fulfil the above criteria.

C&EN @cenmag 22 Aug 2016

If you choose to accept this mission, we'll see you soon for the big unveil! #ACSPhilly #cen12 ow.ly/BnC8303ql0H

This was drawn in about one minute in response the announcement of the Talented 12 at ACS National Meeting in Philadelphia in 2016. It was later printed as a reader response in a subsequent Chemical and Engineering News edition.

C&EN @cenmag

New in our inbox: a poem about the pesticide chlorpyrifos, believe it or not. ow.ly/xAH130e3ZCe

Lisa Jarvis @lisamjarvis 1 Aug 2017

More reader mail in the form of poems, please.

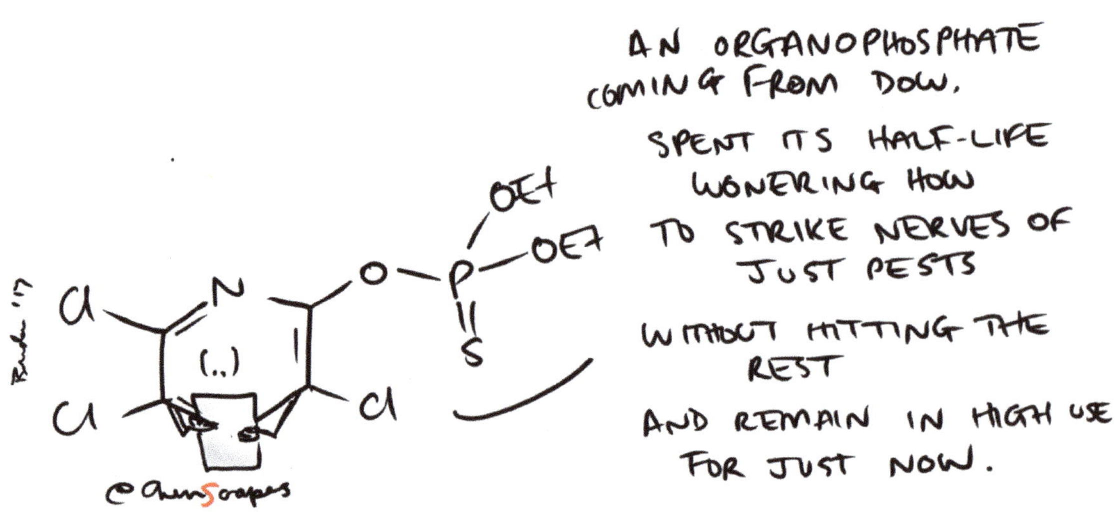

Baran Lab @BaranLabReads 1 Feb 2017

Housanes (available here: bit.ly/2eBnbxE) and much more. Relieve your strain (free download+900 page SI): bit.ly/2kdJsV1

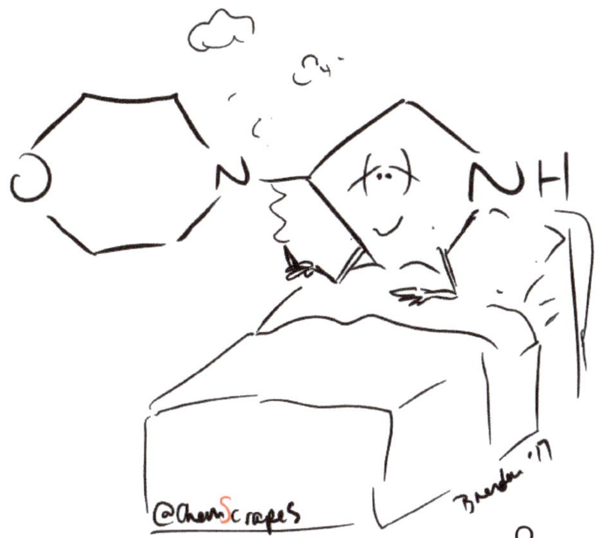

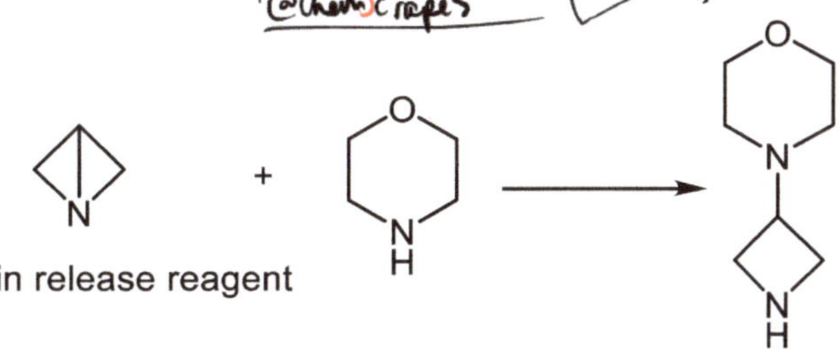

Strain release reagent

Bethany Halford @beth_halford 14 Apr 2017

Baran takes on boron: New work from @BaranLabReads in @sciencemagazine. Read about it in @cenmag cen.acs.org/articles/95/i1...

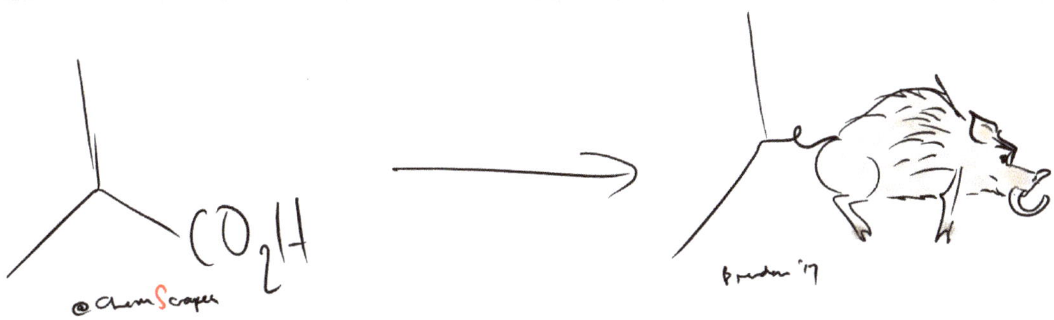

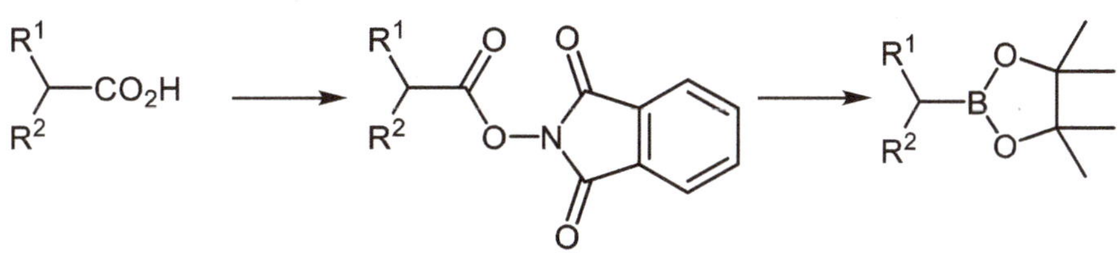

Carboxylic acid Redox active ester Boronic acid pinacol ester

The nice thing about chemistry Twitter is that many of the journal and magazine editors and reporters are part of the community and very active. Every now and then new Journal suggestions pop up which provides opportunities for some fun puns. Below is a contribution to a very long, punny thread about cows; right is response to a challenge by @AlexFGoldberg's suggestion for a new pizza journal; overleaf is an unsolicited suggestion for a new journal based on a particularly colourful thread where @AlexFGoldberg and then ACS president Allison Campbell (@AllisonChem) were comparing socks. All good fun!

nature pizza
GRATE EXPECTATIONS
LIPID COMPOSITION AND
CHEESE TEXTURE
pg 62
REDEFINING Pi
pg 76.
PIZZA-ELECTRICS
THE NEXT BIG WAVE IN NEW MATERIALS

J. AMER. CHEM. SOCK.
Vol. 1. 2017
FEETURE ARTICLE: TRIPODAL LEGANDS FOR SELF HEELING APPLICATIONS
GOLDBERG, CAMPBELL & WANG
@ChemSocks

C&EN @cenmag 10 Jul 2017

On the cover: this project aims to build the human genome from scratch. It isn't going to be easy. ow.ly/kGO930dv6j3

The image below was drawn in response the many instances of pentavalent carbon atoms which appear in chemical adverts and occasionally as errors in magazines. This person is from the Pentavalent Carbon Branch (hence PVC on his hat).

American Chemical Society @AmerChemSociety · 1 Sep 2017

Microbes Could Turn Astronaut Pee into Plastic and Nutrients (Video) ow.ly/uUTs30eJPd6 -- @SPACEdotcom reports on research from #ACSDC

C&EN @cenmag 22 Feb 2017

Spider silk gear is hoping to be the next big thing. But manufacturers are encountering a web of challenges. ow.ly/PJdU309dkrT

Nature News & Comment @NatureNews 21 Apr 2017

Chemists to race single molecules along a 100 nanometre track to drive enthusiasm for nanotechnology: go.nature.com/2pQMdwX

Chemistry World @ChemistryWorld 28 Jul 2016

New kind of antibiotic picked out of the human nose rsc.li/2adLU76

Brendan @ChemScrapes 29 Jul 2016

Replying to @ChemistryWorld

Introducing big pharma's new antibiotics discovery platform.

This is one of those moments where I wish that it was possible to quote a previous Tweet while replying to another AND add a new picture. Put together, this string makes a lot of sense. Ten months apart however, replying to Tweets from different magazines, the connection to the guys picking their noses in the lab scene below seems a little out of the blue.

C&EN @cenmag 18 Apr 2017

Roadkill just might help scientists discover new drugs. ow.ly/i41g30aVLR4 #newscripts

Brendan @ChemScrapes 18 Apr 2017

Replying to @cenmag

Meanwhile at big pharma HQ

Dave Palmer @DavePalmerUSask 11 Apr 2017

Now I'm picturing a pamphlet, showing someone with their face stuck in Büchner, with the title "Is Theoretical Chemistry Right for You?" 😁

> **HartreeFocker** @edsherer
> Replying to @DavePalmerUSask
>
> Student in our lab got too close to suction flask and got face stuck to it during synthesis

Compound Interest @compoundchem 12 Apr 2017

Replying to @DavePalmerUSask and @Chemjobber

Can @ChemScrapes make this image happen?

Chemjobber @Chemjobber 12 Apr 2017

Replying to @compoundchem and 2 others

Pshaw, if I asked for a hybrid-carbon monoxide/elephant riding a unicycle while arm-wrestling E.J. Corey, I know @ChemScrapes could draw it.

3 1 11

Nature News & Comment @NatureNews 5 Feb

Meet the tiny 'robot' whose movements are controlled by magnets. It can walk, jump, roll and even swim - and it could have a range of medical applications.

Replying to @NatureNews

but can it win in Robot Wars?

J Phys Chem @JPhysChem 6 Aug 2017

#WeekendRead #ACSEditorsChoice #OA How Large is the Elephant in the DFT Room? #CompChem #JPCA Jensen @StudyAarhusUni pubs.acs.org/doi/full/10.10...

Nature News & Comment @NatureNews 16 Jul 2017

There should be more public debate around sex robots: go.nature.com/2urHyXQ #NaturesTake

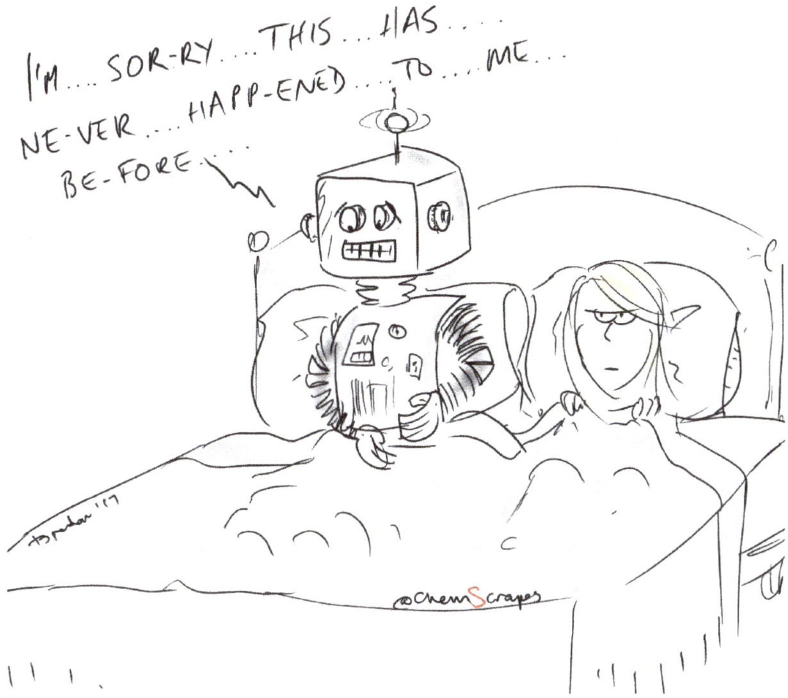

Reactions @ACSReactions 29 Nov

Tis the season for hand sanitizer! Here's how the stuff works, and the truth behind killing "99.99% of all germs." bit.ly/2kawpYr

Tien Nguyen @MustLoveScience 20 Mar
Replying to @MustLoveScience
'Edible underwear is essentially like fruit roll-ups cut into underwear shape' - chem of sex session continues #acsnola

Tien Nguyen @MustLoveScience 20 Mar
Replying to @MustLoveScience
'Because of that simple scrotal support, it's now a class one,' referring to an edible g-string. File that under things I never thought I'd hear at an ACS meeting #acsnola

The end

Sources for chemical structures:

Compound Interest: www.compoundchem.com
ChemSpider: www.chemspider.com
Nicolaou, K. C.; Montagnon, T. *Molecules That Changed the World*. **2008**. Wiley-VCH. Weinheim.

www.chemscrapes.net

CPSIA information can be obtained
at www.ICGtesting.com
Printed in the USA
LVHW070301070120
642630LV00012B/693/P